Abbreviations

ASBo	Archivio di Stato di Bologna
- Prefettura	del dipartimento del Reno
ASMi	Archivio di Stato di Milano
- Popolazione p.a.	Atti di Governo, Popolazione p.a.
- Popolazione p.m.	Atti di Governo, Popolazione p.m.
- Registri	di verbali e di atti di governo e deliberazioni
ASPa	Archivio di Stato di Padova
ASVe	Archivio di Stato di Venezia
- Chioggia	Viceprefettura Chioggia
- Prefettura	Prefettura del dipartimento dell'Adriatico
- San Donà	Viceprefettura San Donà (poi Portogruaro)
ASVi	Archivio di Stato di Vicenza
- Commissariato	Commissariato di polizia
Assemblee	Assemblee della Repubblica Cisalpina
Bertoliana	Biblioteca Civica Bertoliana
- Torre	Archivio storico del Comune di Vicenza (archivio Torre)
Bollettino	See Bibliography, Published Legislative Sources
Iuniori	See Bibliography, Published Parliamentary Sources
Leggi	See Bibliography, Published Legislative Sources
Raccolta	See Bibliography, Published Legislative Sources
Redattore	See Bibliography, Published Parliamentary Sources
Seniori	See Bibliography, Published Parliamentary Sources

Cultures of Identification in Napoleonic Italy, c.1800–1814

Through the lens of identification procedures, this book examines how the processes of state-building affected European societies during the Napoleonic period. By focusing on the Kingdom of Italy, the author shows how the top-down change usually associated with Napoleonic state-building had to compete and share spaces with the agencies of other often-neglected actors such as local bureaucrats, the clergy, and common people.

What emerges is the coexistence of different understandings of personal identities, defined as "cultures of identification". One was rooted in the traditional habits of the population and based on a continuous performance of identities, allowing for a certain degree of fluidity. The other, promoted by the Napoleonic administration, envisaged legal and fixed identities that were to be managed directly by agents of the state. Personal identification in Napoleonic Italy was thus more of a battleground than a mere field of action for the "modernizing" activities of state authorities.

Analysing a period of momentous change for European societies, *Cultures of Identification* can be profitably read by students and researchers interested in the history of state-building, policing, social control, and personal identification.

Stefano Poggi is a research associate at the Austrian Academy of Sciences (ÖAW) of Vienna. After receiving his PhD from the European University Institute, he was a post-doctoral researcher at the University of Padua. His research is mainly concerned with the history of state-building in the long nineteenth century, with a specific focus on social practices and identity issues.

Routledge Studies in the Modern History of Italy

Edited by Carlotta Ferrara degli Uberti (University College London)

Marco Mondini (University of Padua and Italian German Historical Institute-FBK Trent)

Silvana Patriarca (Fordham University)

Guri Schwarz (University of Genoa)

The history of modern Italy from the late 18th to the 21st centuries offers a wealth of dramatic changes amidst important continuities. From occupying a semi-peripheral location in the European Mediterranean to becoming one of the major economies of the continent, the Peninsula has experienced major transformations while also facing continuing structural challenges. Social and regional conflicts, revolts and revolutions, regime changes, world wars and military defeats have defined its turbulent political history, while changing identities and social movements have intersected with the weight of family and other structures in new international environments.

The series focuses on the publication of original research monographs, from both established academics and junior researchers. It is intended as an instrument to promote fresh perspectives and as bridge, connecting scholarly traditions within and outside Italy. Occasionally, it may also publish edited volumes. The sole criteria for selection will be intellectual rigour and the innovative character of the books.

It will cover a broad range of themes and methods – ranging from political to cultural to socio-economic history – with the aim of becoming a reference point for groundbreaking scholarship covering Italian history from the Napoleonic era to the present.

Conscription and the Military from 1814 to 1914
Marco Rovinello

Jewish Displaced Persons in Italy 1943–1951
Politics, Rehabilitation, Identity
Chiara Renzo

Fathers of the Lega
Populist Regionalism and Populist Nationalism in Historical Perspective
George Newth

Cultures of Identification in Napoleonic Italy, c.1800–1814
Stefano Poggi

Cultures of Identification in Napoleonic Italy, c.1800–1814

Stefano Poggi

Routledge
Taylor & Francis Group
LONDON AND NEW YORK

First published 2024
by Routledge
4 Park Square, Milton Park, Abingdon, Oxon OX14 4RN

and by Routledge
605 Third Avenue, New York, NY 10158

Routledge is an imprint of the Taylor & Francis Group, an informa business

British Library Cataloguing-in-Publication Data
A catalogue record for this book is available from the British Library

ISBN: 978-1-032-53513-5 (hbk)
ISBN: 978-1-032-53514-2 (pbk)
ISBN: 978-1-003-41243-4 (ebk)

DOI: 10.4324/9781003412434

Typeset in Sabon
by codeMantra

Quand vint l'heure de la sortie du bagne, quand Jean Valjean entendit à son oreille ce mot étrange: tu es libre! le moment fut invraisemblable et inouï, un rayon de vive lumière, un rayon de la vraie lumière des vivants pénétra subitement en lui. Mais ce rayon ne tarda point à pâlir. Jean Valjean avait été ébloui de l'idée de la liberté. Il avait cru à une vie nouvelle. Il vit bien vite ce que c'était qu'une liberté à laquelle on donne un passe-port jaune.

Victor Hugo, Les Misérables

The proofreading of this book was funded by the European University Institute and the DISSGeA department of the University of Padua. This book is based on the dissertation *A Conflict of Identities: Personal Identification in Revolutionary and Napoleonic Italy (1796-1814)* defended by Stefano Poggi at the Eui on 4 June 2021.

Contents

Acknowledgements

I began working on this subject in September 2016 and submitted the final manuscript of this book in October 2023. Over these seven years, I have received invaluable support both from the academic institutions I have belonged to and from my dear friends.

At the European University Institute, Lucy Riall and Pieter Judson guided my research and mentored my academic journey. After all these years, I am proud to now call them my friends. In Florence, I also benefited from the advice and support of Stéphane Van Damme, Regina Grafe, and Ann Thompson. Throughout the various stages of my PhD research, I had the privilege of discussing my topic with several professors and researchers. Among them, I would like to extend special thanks to Livio Antonielli, Alessandro Buono, Simona Cerutti, Vincent Denis, Dominique Kalifa, Stefano Levati, Chiara Lucrezio Monticelli, and Carmine Pinto. Upon returning to Padua for a post-doctoral fellowship, I continued these discussions with several colleagues who had once been my professors. I particularly appreciate the contributions of Enrico Francia, who served as my post-doctoral mentor, as well as Giulia Albanese and Alfredo Viggiano. Speaking of Padua, it is impossible not to mention Carlotta Sorba, who has been a constant, reassuring presence, both personally and intellectually, throughout these years.

One of the advantages of an academic life is the opportunity to meet many interesting individuals from diverse backgrounds. I have made numerous friends in Florence, Paris, and Padua. While it would be impossible to list and thank them all, I want to acknowledge the special support and friendship of Ignacio García de Paso, Marco Cozzani, and Daniel Banks.

The EUI and the DISSGeA department at the University of Padua have provided substantial support in recent years. I extend my gratitude to all the administrative personnel who assisted me with my activities. Furthermore, I would like to thank both institutions for funding the proofreading of my English, carried out by Daniel Banks, to whom I must express my gratitude twice. I must also thank my friend Louis Le Douarin for the maps contained in this book.

Finally, for my family, friends, and partners whom I have not mentioned here, no public acknowledgement would suffice. For what it is worth, thank you.

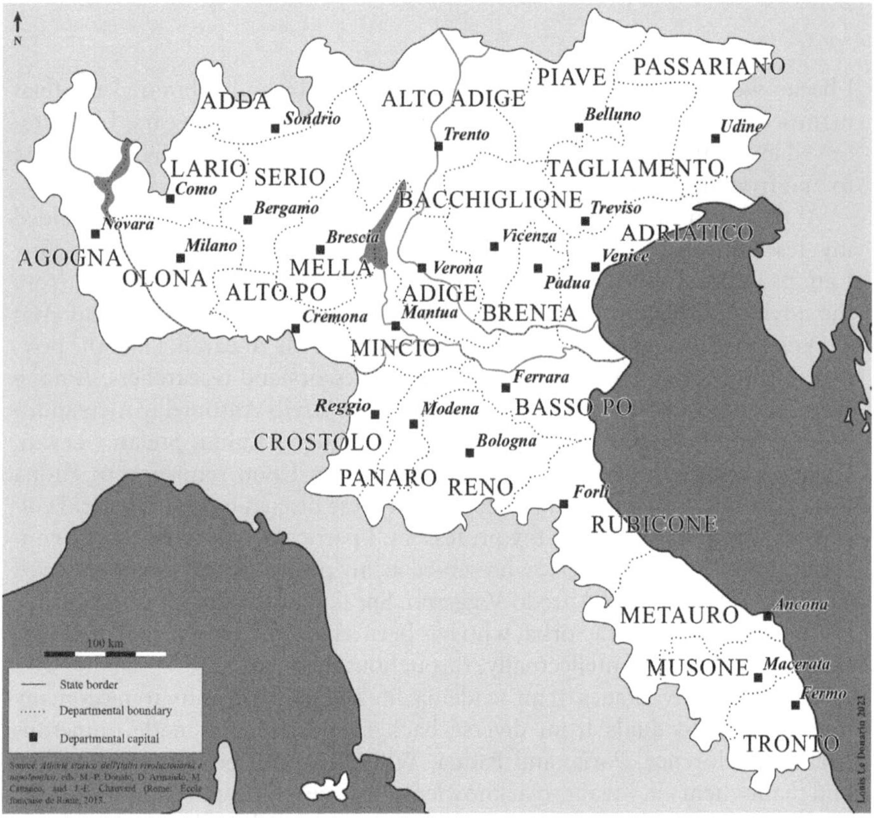

Figure I Administrative map of the Kingdom of Italy in 1810.

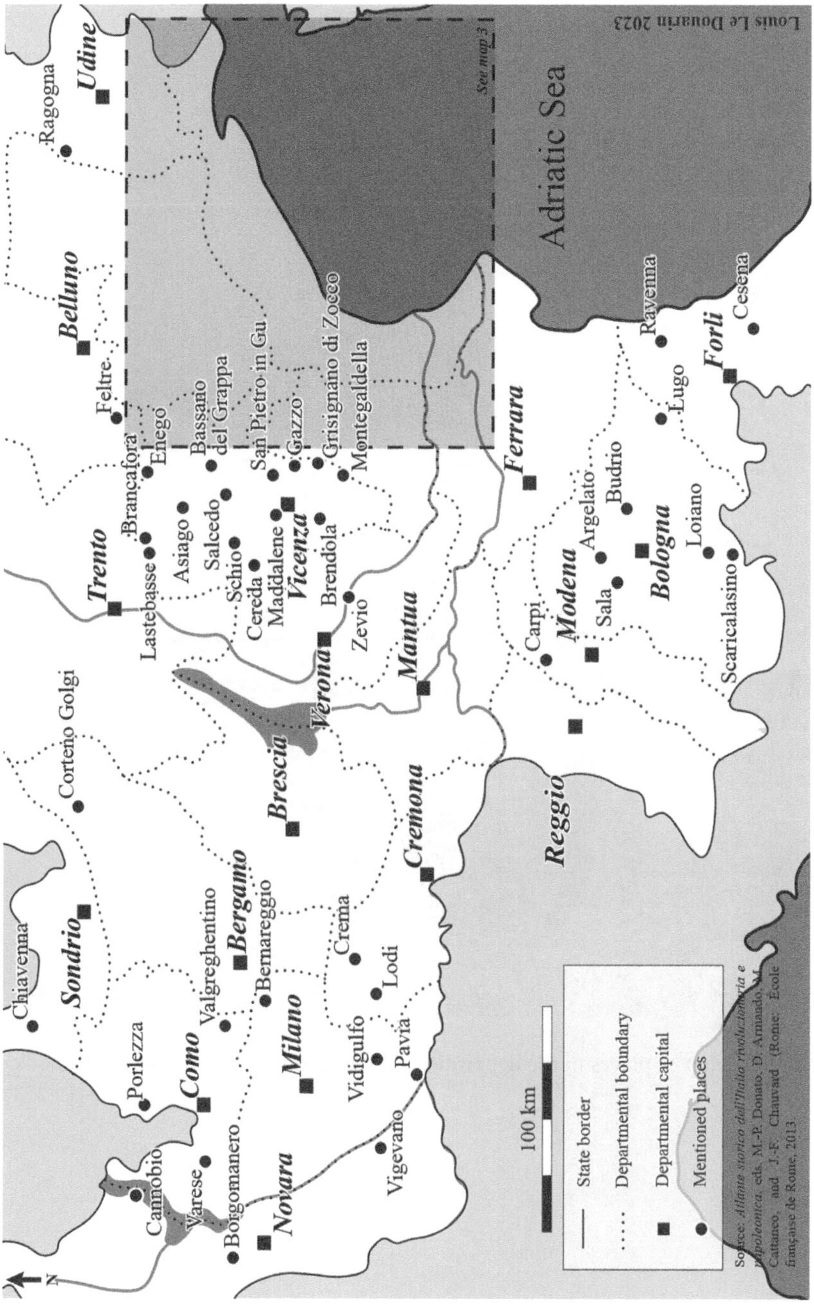

Figure II Map of places mentioned in the book.

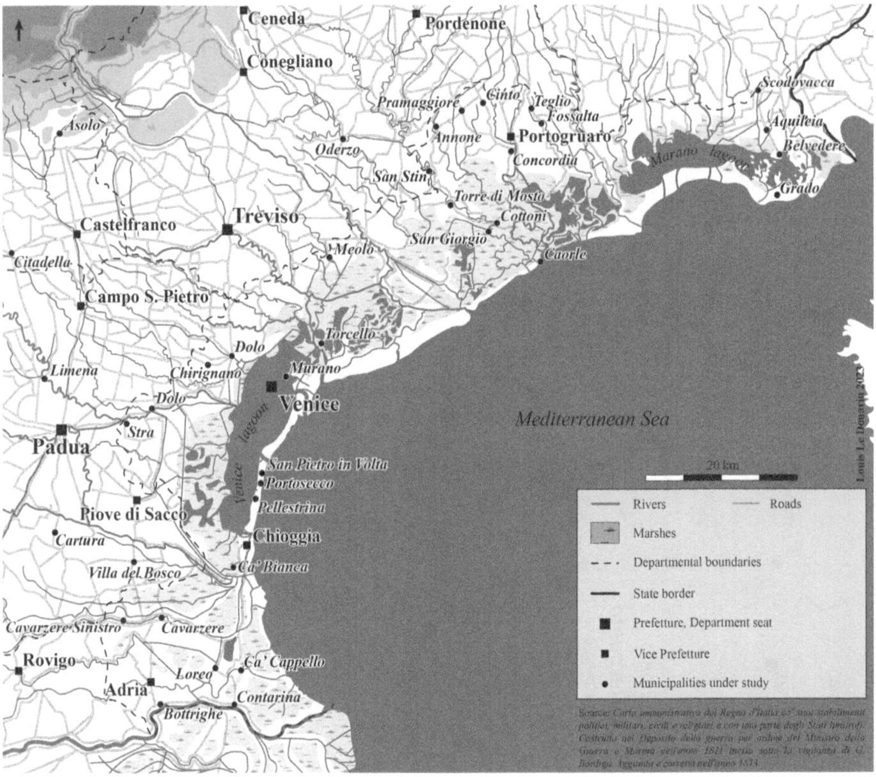

Figure III Map of places in the department of the Adriatico mentioned in the book.

Introduction

On 17 September 1810, in Milan, Antonio Trentin appeared naked in front of a medical committee appointed by the order of the Council of ministers of the Kingdom of Italy. Four of the most prestigious Italian physicians convened to verify whether the young organist was a man or a woman. Two months earlier, Trentin and his fiancée had tried to proceed with their marriage banns in the small village of Cartura, some 340 kilometres away from the capital. Since 1806, according to the new Napoleonic legislation, religious marriage ceremonies had to be preceded by a civil marriage celebrated in front of a municipal registrar. The registrar in Cartura was aware of the peculiar conformation of Trentin's genitals, and had thus decided to block the procedure and report the situation to the local vice-prefect. From the vice-prefect, the case had passed to the prosecutor of Padua, who in turn had called the attention of the minister of Justice to the matter. Trentin had then been summoned to the capital, where his body was put under the scrutiny of the government-appointed medical committee. Two of the committee members – Pietro Moscati and Giovanni Rasori – were known to be involved in a harsh personal and scientific dispute with each other, but they nevertheless agreed in saying that there was no doubt that Trentin was a man – and that he could thus marry. The young couple ultimately married in January 1811, first with a civil ceremony and then, a week later, with a religious one celebrated by the parish priest of Cartura.[1]

This situation would not have been conceivable a few years previously. The civil registration of marriages had in fact been introduced by the pro-French governments in Italy during the previous decade, building on the post-revolutionary administrative tradition of the 1790s. This was just one of the many initiatives Napoleonic authorities adopted in the field of state-building throughout Europe. Over the course of a few short years, among many other things, the pro-French governments built uniform administrative systems that supplanted traditional forms of local autonomy; established a yearly conscription that drafted hundreds of thousands of young men; and reformed the legal system by introducing several new codes of laws.[2] People in Europe reacted in several different ways to these governmental actions, with attitudes

DOI: 10.4324/9781003412434-1

that went from open resistance to cooperation.[3] In several cases, the aims of these reforms were far too ambitious for the Napoleonic states to actually accomplish. Nevertheless, these governmental actions deeply changed the role that the state and its agents played in European societies.

This was also true for the ways in which Napoleonic Europe defined personal identities. Deciding whether a person was a man or a woman, as in the case of Antonio Trentin, had not been a task usually performed by state authorities in previous centuries. Similarly, state agents had remained marginal in the definition of other aspects of the personal identities of their subjects, such as their names, their nationalities, their places of residence, and their age.[4] These aspects were at the centre of several reforms that the Napoleonic governments tried to implement all over Europe.[5] The effects of these governmental initiatives in the field of personal identification are the main subject of this book. In my analysis, I will focus on the peculiar context of Napoleonic Italy, where – as I argue further on – an administrative laboratory emerged from the combination of the political legacy of the French revolution with the legacy of eighteenth-century Habsburg reformism. Narrowing the scope of analysis to this particular context will highlight some of the distinctive features of the Napoleonic attempt to re-shape European societies.

In so doing, I will adopt a plural research perspective that does not just consider the agency of central governments, but also the agencies of other crucial actors, such as local bureaucrats, the clergy, and the common people. What emerges from this multi-faceted analysis is a scenario in which the phenomena I analyse are much more than a set of changes imposed from above on a 'passive-aggressive' society by a modernising government. To grasp the complexity of this scenario, I introduce the notion of "cultures of identification", which I will define below. Using this concept, this book tries to understand not just the rationale that drove the government's initiatives, but also the considerations that inspired the agency of local bureaucrats and ordinary people.

The history of Napoleonic Europe – and of Napoleonic Italy in particular – is often reduced to a series of more or less successful innovations introduced in the field of state-building. This book does not deny the extent and the importance of these changes, whose impact was so great that they were largely upheld by European state institutions well after the Restoration of 1814–1815.[6] However, this book aims to make the picture more complex by introducing neglected actors and considering their actions not merely as a reaction to state initiatives, but as part of their own different and consistent agendas.

Napoleonic Italy as an Administrative Laboratory

In this book, I focus on a specific area of what is commonly referred to as 'Napoleonic Italy': the central and northern territories of the Italian peninsula that were governed first by the Cisalpine Republic, then by the Italian republic, and finally by the Kingdom of Italy (Figure I). Even if the monarchic period

(1805–1814) will be at the core of my analysis, I will also consider previous republican experiences (1796–1805), especially in the part dedicated to government planning. The choice of this specific part of the Italian Peninsula will allow me to consider a state which had a peculiar position within the Napoleonic system. The Kingdom of Italy was the Italian territory in which French influence was most stable and long-lasting. The Lombard core of the kingdom had been part of the French system since 1796 and remained so until the end of the Napoleonic era, with the exception of a brief period of Habsburg occupation in 1799–1800. At the same time, this core had been part of the Habsburg Empire for much of the eighteenth century. In this context, the reformist initiatives promoted by the governments of Maria Theresa and Joseph II had shaped an administrative tradition that deeply influenced the action of the pro-French élites.[7]

Thanks to their relative political stability, the Cisalpine/Italian Napoleonic Governments were able to develop autonomous political, legislative, and administrative initiatives under the protection of the French armies. The fact that this was effectively a "puppet-state" granted the Italian élites a certain degree of autonomy, as they creatively combined French influence with local eighteenth-century reformist traditions. The government – at both the local and central levels – was almost entirely composed of Italians – a feature that set these territories apart from the kingdom of Naples and the areas of the peninsula that were directly annexed to the empire.[8] This last fact evocatively points to the double nature of the Italian state. On the one hand, it was an autonomous polity with an independent bureaucracy, government, and institutions. On the other, its dependency on the French empire made it more like a protectorate than an independent state.[9] A focus on Italian administrative initiatives can thus shine a light on the different degrees to which Napoleonic rule adapted to the specificities of the different territories that depended on it. In this sense, the Italian context can be seen as a peculiar administrative laboratory in which different political cultures came together in an original combination.

At the same time, this administrative initiative had to confront a set of extremely diverse territories and social realities. At the peak of its extension, the Kingdom of Italy covered most of the Po plain – from the Sesia river to the Julian Pre-Alps. Its territory also included vast mountainous areas, from the Alps to the Apennines. Its eastern border was determined by the long Adriatic coastline. The kingdom's territory included a multitude of different landscapes and climates. Within its borders, a traveller could go from the cereal fields of agricultural Lombardy to the remote and majestic mountains of the Cadore, passing through the marshlands of the Venice lagoon along the way. This geographical variety was mirrored by a variety in forms of human presence, from the small and remote villages of the Alpine valleys to the well-connected cities of Milan and Venice, each of which was home to over one hundred thousand inhabitants. Connections between the different parts of the kingdom was assured by an extensive road network, which could, however, be disrupted for several months at a time by the harsh winter weather that affected large parts of the Italian territory.

By 1809, around 6,700,000 people lived in the Kingdom of Italy – one-third of the entire population of the peninsula. Most of these inhabitants (between 70 and 85 percent) lived in the countryside. Here, peasants and farmers were experiencing a process of pauperisation that had started in the mid-seventeenth century and had been accelerated by eighteenth-century reforms such as the privatisation of commons, new taxes, and the weakening of religious orders. They were also affected by population growth and the worsening conditions of farming contracts. The situation was different in urban centres, with their much more varied social makeup. Here, most of the active population (between 55 and 70 percent) was occupied in artisanal or domestic work. Both in the cities and in the countryside, the presence of a growing stratum of beggars and vagabonds unnerved the élites.[10] The severe reduction in the number of charitable religious organisations had deprived the poorer part of the population of substantial means of subsistence. The blows of reformist and Napoleonic restructurings further thinned the ranks of the Italian clergy: in some territories, such as the diocese of Brescia, the number of clergymen was halved between 1765 and 1810.[11] Various expropriations and privatisations contributed to the formation and the strengthening of a class of landowners – both noble and bourgeois – who controlled the three main economic resources of the Italian economy: wheat, silk, and wine.[12]

These landowners formed the backbone of the pro-French élite. Napoleon promoted a process of *ralliement* among the different sections of the Italian ruling class.[13] Nobles, landowners, and intellectuals were asked to cooperate in the management of the state via the creation of an increasingly expanded bureaucracy. At the level of the central state, this policy was embodied by the creation of three "electoral colleges" in 1802: one for landowners, one for merchants, and one for intellectuals. In this way, the Napoleonic state tried to involve economic and intellectual élites in the management of political power. In so doing, it reproduced the fundamental traditional structures of power that had sustained the pre-Napoleonic states. In particular, it kept alive the polycentric nature of Italian politics, while trying to integrate local élites into the framework of the new administrative state. As Michael Broers has argued, in Italy, Napoleonic institutions had to deal with territories characterised by a deeply rooted political cleavage between urban centres and their rural and mountainous peripheries.[14]

Napoleonic states tried to tackle this situation by developing a dense administrative network. At the local level, they replaced traditional forms of autonomy with municipalities that ultimately constituted a lower level of central state organisation. These municipalities were made up of officials who had been appointed from above.[15] The state grouped municipalities into districts and districts into departments. Each departmental administration was headed by a prefect chosen by the central government. The prefect was generally a career bureaucrat and, in most cases, came from a different region of the kingdom to the one he was posted to. In the smaller centres of each department, vice-prefects exercised power on the prefect's behalf. The prefectures and vice-prefectures were – in the areas they controlled – the supreme representatives of local state power. The prefects communicated directly with the central government, and, in particular, with the minister of the Interior.[16] Each department also had an ourt of First Instance in the departmental seat, presided over by a prosecutor-general who answered directly to the minister of Justice. At a lower level, justices of the peace were charged with minor judicial duties.[17] In most municipalities, the mayor was in charge of police functions, while police commissioners were appointed for larger cities. In addition to the police, a gendarmerie was established in 1803 with the specific aim of monitoring the countryside and tackling desertion and banditry.[18]

Over time, central power was increasingly wielded by the state's executive bodies. During the first Cisalpine Republic (1797–1799), the executive branch was composed of a five-member directory, who also appointed ministers. The directory's authority, however, suffered from frequent French interventions: between 1798 and 1799, the executive's composition was changed five times after five different *coups-d'état* promoted by the French military authorities – two against the moderates and three against the radicals. Legislative power was in the hands of a Great Council (renamed Council of Juniors in April 1798), that held the legislative initiative, and a Council of Seniors, that could approve or reject proposals from the lower chamber. The members of the two councils were always, in practice, appointed by the French military authorities, who often intervened in the composition of these chambers. When the Cisalpine Republic was reinstated in 1800, after a brief period of Habsburg occupation, legislative power was given to a Legislative Council, while executive power was wielded by an Executive Committee. Both were appointed by the French First Consul, Napoleon.

The creation of the Italian Republic in 1802 once again changed the way central power was organised. The executive branch gained pre-eminence over the legislative body. The president of the republic – Napoleon himself – was nominally the repository of executive power, and could appoint the other central bodies. He also appointed the vice-president, Francesco Melzi d'Eril, who was, in practice, in charge of most of the republic's day-to-day executive duties.[19] The president held the legislative initiative: he could submit proposals to a Legislative Council appointed by himself. In order to be fully approved, laws had to be ratified by a larger legislative body, elected by the aforementioned electoral colleges of landowners, merchants, and intellectuals. The monarchic turn of 1805 changed the highest levels of the Italian state for good. Napoleon became king, while his stepson Eugène de Beauharnais became viceroy.[20] The Legislative Council was integrated with the Council of state, the central body of the state, which included all of the kingdom's top officials. Its role, however, remained mostly consultative.[21]

As power shifted from the legislative to the executive, the ministers became the Italian government's most important actors. They were not just in charge of the implementation of laws and regulations, they also promoted most of the state's legislative initiatives. With the creation of the Italian republic in 1802, they came to be at the centre of the political debates that led to the proposal of draft laws and provisions. Even if the head of state – or his deputy – still had a final say on everything, the ministers occupied spaces that were left open by the decreasing power of the legislative bodies. Their action, however, was often limited by the fact that ministerial jobs were highly unstable. The minister of the Interior, for instance, changed five times between 1800 and 1809. But other ministerial posts were more stable. The minister of Justice only changed once between 1802 and the end of Napoleonic rule.

This complex administrative and political structure was engaged for almost two decades in an ambitious programme of reforms. A new conception of the state, derived both from post-revolutionary legacies and eighteenth-century reformist traditions, slowly tried to become a reality. This ambitious state initiative interfered with the very roots of the identities of subjects.

Cultures of Identification

Something substantial changed in how people were identified during the Age of Revolutions. In previous centuries, state authorities had only marginally bothered themselves with identification procedures. However, starting in the eighteenth century, European states had begun to elaborate new practices in this field. As Vincent Denis has shown, the kingdom of France was at the forefront of these innovations. It established new means of control over mobile elements of the population, army members, and foreigners. The turning point in this slow process of accumulation of administrative traditions was the 1789 revolution. The latter brought about a systematisation of these eighteenth-century innovations, along with their extension to a national level, leading to the rise of a 'new identification regime'.[22] According to Caplan and Torpey, the development of large-scale forms of identification was a natural consequence of the type of national citizenship that the revolutionary authorities were creating.[23] However, phenomena of this kind were not limited to areas that fell under the influence of the revolution. In Habsburg Venetia (1799–1804), for instance, the administrative state developed a similar sort of "paper citizenship" to achieve a proper "incorporation" of society, even if it did not have the necessary means.[24]

These processes had long-lasting consequences in nineteenth-century Europe. According to Gérard Noiriel, nineteenth-century France underwent a full-blown '*révolution identitaire*'.[25] The establishment of a system of civil registration provided the state with a way of reaching millions of individuals.[26] Noiriel also highlights the role registration procedures played in the general integration of the population into the French national project: the mandatory use of French in civil acts, along with the francisation of names and surnames, played an important part in processes of national assimilation.[27] At the same time, the establishment of new identity documents – and of passports in particular – brought about a materialisation of the sense of belonging to a French national community.[28] Edward Higgs has noted how, in the nineteenth century, personal identification was increasingly managed by central bureaucracies by way of databases. Identification processes moved away from single individuals, following the parallel estrangement of public power from the population.[29]

The historiography has thus stressed the close connection between changes to personal identification in nineteenth-century Europe and the development of nation-states.[30] On the one hand, administrative states wanted (and needed) tools with which to control their population and the identities of their citizens. On the other, the idea of belonging to a national community implied a need for a means of formalising and certifying a uniform sort of citizenship for the whole population of a specific state.[31] The changes in identification systems, in this sense, resulted from political projects emerging from the new nature of the state.

These same scholars have, in contrast, also emphasised changes in society that took place in and before the nineteenth century. Caplan and Torpey, for instance, mention the rise of mobility during the previous century.[32] Increased interconnection has also been seen as a key factor in these changes. According to Noiriel, the new identification procedures also responded to the needs of the new civil society itself. Passports and identity cards offered guarantees in a society that was becoming increasingly interconnected. In Noiriel's opinion, these bottom-up requests merged with the will of the revolutionary state to create a new nationhood based on the equality of rights and duties.[33] The English case, on the other hand, seems to contradict these sociological interpretations of the changes that took place during the nineteenth century. Edward Higgs has for instance pointed out that, in England, the centralisation of identification only happened in the second-half of the nineteenth century, while the previous intense period of industrialisation, mobility, and urbanisation was marked by a decreasing role of the state in the realm of personal identification.[34]

Despite the different chronologies and modalities in which European states intervened in the realm of identification, they were all inspired by a common understanding of what the categories of personal identification were and how they were to work. Starting with the Age of the Revolutions and throughout the nineteenth century, state authorities promoted a common culture of identification throughout the European continent.[35] First of all, this culture entailed the creation of stable and distinct individual identities that implied rights and obligations vis-à-vis state authorities.[36] Every citizen had to have a stable gender, name, surname, birthplace, and age. The legal identity composed by the sum of these elements could only be changed by resorting to the state authorities themselves. These identities had to be certified and validated directly by state agents, via procedures and certificates that were established by law. According to John Torpey, once they had assured their predominance over the means of movement, states wanted to 'develop the capacity to "embrace" their own citizens in order to extract from them the resources they need[ed] to reproduce themselves over time'.[37] In this sense, state cultures of identification tended to minimise the role of communities and social networks in the definition of individuals' identities. This last feature is an indication of the "universal" value of state identities, as these were meant to be valid across entire national territories. The manufacturing of these legal identities was part of a much larger process by which European states tried to create "legible societies" between the eighteenth and the nineteenth centuries.[38] In this sense, the state culture of identification was linked to a will to establish new means of control over populations and territories.[39]

State authorities, however, did not introduce their new culture of identification into a vacuum. In fact, over the previous centuries, different European populations had developed a plurality of practices aimed at certifying the identity of individuals.[40] These social practices, in turn, expressed a different culture of identification to the new, state-driven one. As Alessando Buono has shown, in *ancien régime* societies, personal identity was something that had to be constantly claimed through use. In other words, identity was defined more by public acts than by specific documents.[41] The definition of personal identities was the product of a continuous process of "contextualisation" in which individuals performed a certain identity that their communities could in turn accept or refuse, depending on each situation.[42] This meant that individuals enjoyed a certain degree of fluidity in the various aspects that composed their identity, such as name, nationality, and even gender.[43] The communal culture of identification expressed by these practices entailed a conception of personal identity that was highly horizontal, performative, and situational. But this conception was ultimately not compatible with increasing push by state authorities to obtain knowledge of their populations.

What happened in nineteenth-century Europe is that different cultures of identification started to compete for social hegemony. One of these cultures, rooted in people's customs, implied fluid and communal identities. The other, promoted by state authorities, entailed legal and stable identities certified by agents of the state itself. This book analyses this "conflict of identities" in its early stage, combining the study of these different cultures of identification with an analysis of the agencies of the various players involved in processes of identification, from national politicians to common Italians.

My analysis is centred on the study of two main pillars of nineteenth-century personal identification: identity documents and the civil registers of births, marriages, and deaths. I analyse these tools of identification in the context of Napoleonic Italy, a peculiar laboratory in which post-revolutionary administrative cultures met, clashed, and occasionally even teamed up with longstanding customary practices. It was in this context that a variety of players used these state-driven processes to create new opportunities and meanings for themselves, crafting a reality that considerably differed from the original plans of state élites.

Notes

1 For a reconstruction of the case: Stefano Poggi, "'Non si tenti un matrimonio fra due donne'. 'Vero sesso' e impotenza nell'Italia napoleonica," in *Alla prova delle passioni. Sessualità non conformi e soggettività fra età moderna e contemporanea*, ed. Tommaso Scaramella (Pisa: Edizioni ETS, 2024).
2 Stuart Woolf, *Napoleon's Integration of Europe* (London-New York: Routledge, 1991); Aleaxander Grab, *Napoleon and the Transformation of Europe* (Bagingstroke-New York: Palgrave MacMillan, 2003).
3 See the various chapters of *Collaboration and Resistance in Napoleonic Europe: State-Formation in an Age of Upheaval, c. 1800–1815*, ed. Michael Rowe (Bagingstroke-New York: Palgrave Macmillan, 2003), in particular 1–18. Italian historiography has focused, in particular, on the forms of rebellion known as *insorgenze*: see, for instance, Anna Maria Rao, *Folle controrivoluzionarie. Le insorgenze popolari nell'Italia giacobina e napoleonica* (Roma: Carocci, 1999).
4 For an overview of the history of personal identification in early modern Europe: Valentin Groebner, *Who Are You? Identification, Deception and Surveillance in Early Modern Europe* (New York: Zone, 2007).
5 In particular, for marriage legislation: Francis Ronsin, *Le contrat sentimental: débats sur le mariage, l'amour, le divorce, de l'Ancien Régime à la Restauration* (s.l.: Aubier, 1990), 177–234; Yves Bruley, "Mariage et famille sous Napoléon. Le droit entre religion et laïcité," *Napoleonica* 14 (2012): 111–126.
6 See *Napoleon's Legacy: Problems of Government in Restoration Europe*, eds. Lucy Riall and David Laven (Oxford: Berg, 2000). More specifically regarding personal identification see, for instance, Laura Di Fiore, *Alla frontiera. Confini e documenti di identità nel Mezzogiorno continentale preunitario* (Soveria Mannelli: Rubbettino, 2013).
7 Livio Antonielli, "L'Italia di Napoleone: tra imposizione e assimilazione di modelli istituzionali," in *Gli imperi dopo l'Impero nell'Europa del XIX secolo*, eds. Marco Bellabarba, Brigitte Mazohl, Reinhard Stauber and Marcello Verga (Bologna: il Mulino, 2008), 409–431.
8 Livio Antonielli, *I prefetti dell'Italia napoleonica* (Bologna: il Mulino, 1983); Antonio Liva, *Gli impiegati*, in *La formazione del primo Stato italiano e Milano capitale, 1802–1814*, ed. Adele Robbiati Bianchi (Milano: Istituto lombardo Accademia di scienze e lettere, 2006), 871–882. On the Lombard reformist tradition: Alexander Grab, "Enlightened Despotism and State Building: The Case of Austrian Lombardy," *Austrian History Yearbook* 19/2 (1984): 43–72.
9 Alain Pillepich, *Napoleone e gli italiani* (Bologna: il Mulino, 2005), 46.
10 Francesca Meneghetti Casarin, *I vagabondi. La società e lo Stato nella Repubblica di Venezia alla fine del '700* (Roma: Jouvence, 1984); Simona Cerutti, "Les 'misérables' en droit italien au XVIII e siècle," in *Gens de passage en Méditerranée de l'Antiquité à l'époque moderne. Procédures de contrôle et d'identification*, eds. Claudia Moatti and Wolfgang Kaiser (Paris: Maisonneuve & Larose, 2007), 223–242.
11 Stefano Levati, "Società," in *Atlante storico dell'Italia rivoluzionaria e napoleonica*, eds. Maria Pia Donato, David Armando, Massimo Cattaneo, and Jean-François Chauvard (Rome: École française de Rome, 2013), 123–160.
12 Walter Panciera and Saverio Russo, "Economia," in *Atlante storico dell'Italia rivoluzionaria e napoleonica*, 164–201.
13 Carlo Zaghi, *L'Italia di Napoleone* (Torino: UTET, 1989), 2–6.
14 Michael Broers, *The Napoleonic Empire in Italy, 1796–1814. Cultural Imperialism in a European Context?* (Basingstoke: Palgrave Macmillan, 2005), 11–22.
15 Emanuele Pagano, *Enti locali e stato in Italia sotto Napoleone. Repubblica e Regno d'Italia (1802–1814)* (Roma: Carocci, 2007), 111–118.
16 On the Italian prefects, see: Antonielli, *I prefetti.*

17 *Ibid.*, 437–474; Alfredo Viggiano, "Il disordine delle comunità. I giudici di pace napoleonici nei dipartimenti veneti," in *Forme e pratiche di polizia del territorio nell'Ottocento preunitario* (Soveria Mannelli: Rubbettino, 2011).
18 Simona Mori, *Polizia e statualità nel primo Ottocento: l'esperienza lombardo-veneta e la cultura professionale italiana* (Soveria Mannelli: Rubbettino, 2017).
19 As Livio Antonielli has pointed out, Melzi found a peculiar compromise between the French model and the one inherited from the Habsburg tradition. Livio Antonielli, "L'Italia di Napoleone," 409–431.
20 Carola Oman, *Napoleon's viceroy, Eugène de Beauharnais* (London: Hodder & Stoughton, 1966).
21 Maria Francesca Baldinelli, "Il Consiglio di Stato del Regno d'Italia napoleonico: le procedure e il funzionamento," *Società e storia* 78 (1997): 789–828.
22 Vincent Denis, *Une histoire de l'identité: France, 1715–1815* (Seyssel: Champ Vallon, 2008), 445–450.
23 Jane Caplan and John Torpey, "Introduction," in *Documenting Individual Identity: The Development of State Practices in the Modern World*, eds. Jane Caplan and John Torpey (Princeton, NJ; Oxford: Princeton University Press 2001), 7–8.
24 Alfredo Viggiano, "Carte d'identità nella Venezia Austriaca (1799–1804)," in *Controllare il territorio. Norme, corpi e conflitti tra medioevo e prima guerra mondiale*, eds. Livio Antonielli and Stefano Levati (Soveria Mannelli: Rubbettino, 2013), 363–378; Alfredo Viggiano, "Individuare e classificare. Passaporti concessi e negati nella Venezia austriaca del primo Ottocento," in *Procedure, metodi, strumenti per l'identificazione delle persone e per il controllo del territorio*, ed. Livio Antonielli (Soveria Mannelli: Rubbettino, 2014), 105–131. These initiatives can be considered as the natural consequence of the development of a new form of imperial citizenship starting in the reign of Joseph II: Pieter M. Judson, *The Habsburg Empire: A New History* (Cambridge, MA; London: The Belknap Press of Harvard University Press, 2016), 51–54, 75–78. On the concept of "paper citizenship": Marco Meriggi, "La cittadinanza di carta," *Storica* 16 (2000): 107–120.
25 Gérard Noiriel, "Introduction," in *L'identification. Genèse d'un travail d'Etat*, ed. Gérard Noiriel (Paris: Belin, 2007), 16–20.
26 Gérard Noiriel, "L'identification des citoyens. Naissance de l'état civil républicain," *Genèses* 13 (1993): 17–19.
27 Noiriel, "L'identification des citoyens," 26–27.
28 Gérard Noiriel, "Surveiller les déplacements ou identifier les personnes? Contribution à l'histoire du passeport en France de la I^e^ à la IIIe République," *Genèses* 30 (1998): 78.
29 Edward Higgs, *Identifying the English: A History of Personal Identification 1500 to the Present* (London-New York: Continuum, 2011), 143–170.
30 The historiography on personal identification initially focused precisely on how state authorities identified individuals throughout history: see Caplan and J. Torpey, eds., *Documenting Individual Identity: The Development of State Practices in the Modern World* (Princeton, NJ; Oxford: Princeton University Press, 2001).
31 For instance: Noiriel, "Introduction," 16–20.
32 Caplan and Torpey, "Introduction," 6.
33 Noiriel, "L'identification des citoyens," 26–28. A similar stance can be found in Noiriel, "Surveiller les déplacements".
34 Higgs, *Identifying the English*, 205–210.
35 I define "culture of identification" as a consistent system of understandings, procedures, and practices related to personal identities that can be identified in a specific historical context. This definition is largely inspired by the definition of a "culture of vigilance" proposed by Arndt Brendecke and Paola Molino, "The Cultures of Vigilance. Historicizing the Role of Private Attention in Society: An Introduction," *Storia della Storiografia* 74/2 (2018): 11–16.

36 Caplan and Torpey, "Introduction," 7–8.
37 John Torpey, *The Invention of the Passport: Surveillance, Citizenship and the State* (Cambridge: Cambridge University Press, 2000), 1–2.
38 James C. Scott, *Seeing Like a State: How Certain Schemes to Improve the Human Condition Have Failed* (New Haven, CT; London: Yale University Press, 1998), 12–83.
39 Valentin Groebner has nevertheless highlighted the constant desire of every state to identify its subjects even before the nineteenth century: Valentin Groebner, *Who Are You? Identification, Deception and Surveillance in Early Modern Europe* (New York: Zone, 2007), 205–210.
40 Recent historiography has tended to highlight the role of non-state agents in identification procedures, by also widening the analysis both chronologically and geographically: Keith Breckenridge and Simon Szreter, eds., *Registration and Recognition: Documenting the Person in World History* (Oxford: Oxford University Press, 2012); Ilsen About, James Brown, and Gayle Lonergan, eds., *Identification and Registration Practices in Transnational Perspective: People, Papers and Practices* (Basingstoke: Palgrave Macmillan, 2013).
41 Alessandro Buono, "Tener Persona. Sur l'identité et l'identification dans les sociétés d'Ancien Régime," *Annales HSS* 75/1 (2020): 75–111.
42 Alessandro Buono, "Le procedure di identificazione come procedure di contestualizzazione. Persone e cose nelle cause per eredità vacanti (Stato di Milano, secc. XVI–XVII)," in *Procedure, metodi, strumenti per l'identificazione delle persone e per il controllo del territorio*, ed. Livio Antonielli (Soveria Mannelli: Rubbettino, 2014), 35–65. On the performative nature of personal identity: Erving Goffman, *The Presentation of Self in Everyday Life* (Edinburgh: University of Edinburgh, 1956), 1–46.
43 On the fluidity of names and national identities see, for instance: Teresa Bernardi, "Mobilità e appartenenze multiple a Venezia: il caso di Tarsia alias Laura Malipiero (1630–1660)," *Genesis* 16/2 (2017): 37–59. Similarly, for gender transitions in the eighteenth and nineteenth centuries, see Jen Manion, *Female Husbands: A Trans History* (Cambridge: Cambridge University Press, 2020).

1 Imagining Identities

The Making of a Culture of Identification

In his memoires written in the last decade of the eighteenth century, Giacomo Casanova recalled a significant discussion he had with the mayor of Augsburg in 1757. A few days after his arrival in town, the local authorities had summoned the Venetian adventurer to the city hall. Here, the mayor asked him why he was using the 'fake surname' Seingalt. Casanova unapologetically answered that Seingalt was his 'real surname', because he was its 'author'. Since nobody had ever used that name, he asserted that nobody could question its use. Moreover, being Seingalt did not mean that he could not also be Casanova. The mayor protested that the only surname he could have was his father's one. Casanova retorted, however, that all surnames had been made up by someone at a certain point, exactly as he had done with Seingalt. He thus considered his made-up surname to be completely legitimate. The mayor, impressed by Casanova's arguments and reassured by his connections to a local banker, let him go without further questioning his double denomination.[1]

A few years after Casanova wrote his memoirs, the Cisalpine authorities approved a temporary law to stem a wave of robberies that were affecting the property rights of the republic's citizens. One of the articles stated:

> Anyone who lies, through their attire, about their gender, and anyone who assumes a fake name, or lies about their homeland, will suffer two months' imprisonment as punishment if it is daytime [when caught], four months if it is night-time.[2]

What separates Casanova's account from this Cisalpine law (and the opinions of the mayor of Augsburg) is a fundamentally different interpretation of what constitutes a "real" or a "fake" name. According to the Cisalpine authorities, every individual needed to have a definite and somehow universal identity. According to Casanova, on the contrary, people could use a variety of identities, according to their own will. At the end of the eighteenth century, the cleavage between these two understandings of personal identity became ever more apparent.

The identification culture materialised by the Cisalpine law did not come out of nowhere. As a matter of fact, in the previous decades, a new interest in identification and registration procedures had spread through governments

DOI: 10.4324/9781003412434-2

across Europe, including in the Italian Peninsula. This does not come as a surprise: parish registers, in particular, were used as crucial legal sources in various sorts of proceedings dealing with property rights, marriage, and inheritance. Moreover, they were also the source of one of the main identity documents in use at the time: parish certificates. Thanks to their registers, parish priests could issue baptism certificates that were used in several identification procedures. Parish registers were thus crucial elements for the identification of people in Europe during the *Ancien Régime*. Their public role was undeniable.

Some European states had a long tradition of interventionism in this field. In 1539, the Kingdom of France established the mandatory registration of christenings and burials – 24 years before the *Tametsi* decree, issued during the Council of Trento, imposed the registration of baptisms and marriages on all Catholic territories. In 1667, the *Code Louis*, a large collection of decrees on justice, put the parish registers under the control of French provincial royal judges, the *greffes du bailliage*. It also mandated that a copy of each register be stored in the archives of the *greffes*. As a result of the widespread flaws in the observance of these regulations, in 1736, a new *déclaration* issued detailed guidelines on the writing and keeping of parish registers. This ruling gave the institutions of royal justice a central role in the parish registration system, turning them into the arbiters of the French subjects' civil status. Finally, in 1787, the edict of Versailles – later known as the "edict of Toleration" – granted civil status to non-Catholic subjects. At the same time, it granted Protestant and Jewish subjects the right to register births, marriages, and deaths in their local parish registers.[3] From early on, French governments acknowledged the public nature of parish registers, and thus intervened in their supervision – while still leaving their management to religious personnel.

This was not the case in the Italian states. Here, parish registers remained under the sole authority of the Catholic clergy until the second-half of the eighteenth century. The Duchy of Milan, part of the Habsburg territories, was one of the first states in which this changed. The interventionist policies of the governments of Maria Theresa and her son Joseph inevitably affected parish registration.[4] In 1768, the government asked Lombard bishops to create duplicates of their local parish registers. The district state archives would collect these copies to make up for the widespread dispersal and alteration of parish registers. However, the bishops opposed the proposals, as they were worried that they would eventually have led to a redistribution of registration fees that were vital in supporting the lower clergy.[5] As of 1770, the government required the bishops to collect annual copies of the parish registers in their dioceses.[6] Twelve years later, as a result of Joseph's patents of toleration, non-Catholic individuals were allowed to record births, marriages, and deaths in local parish registers.[7] Finally, in 1784, the approval of new imperial legislation on marriage resulted in the publication of precise directives on the compilation and management of parish registers. This same legislation clearly indicated that marriage was at the same time both a religious sacrament and a civil contract.[8]

These developments did not occur in the neighbouring Republic of Venice, where the government left parish registers in the hands of the clergy. Here, however, as in Lombardy, religious registration was not up to its legal role. In 1787, a Venetian parish priest proposed a plan to reorganise religious registration in the republic's capital. According to the plan, a new civil official would collect all the completed registers as well as those stored in the Venetian parish archives. In the eyes of the plan's initiator, this would have freed priests from the difficult tasks connected with the management of the registers. The reform would also have simplified the procedures by which the city's inhabitants could obtain certificates.

However, the proposal was opposed by one of the main jurists of the republic, Piero Franceschi, who expressed his worry that its implementation would have impinged on the ecclesiastic jurisdiction set by the Council of Trento. For the same reason, in 1795, Franceschi opposed another proposal by the local authorities in Bergamo aimed at creating public archives for the duplicates of parish registers.[9] The Venetian élites were more preoccupied with preserving social and religious peace than with having efficient registration tools. However, demands for change came precisely from those people – parish priests and local authorities – who suffered the most because of the flaws in the parish registration system.

In eighteenth-century Lombardy and Venetia, the management of parish registers was a political issue. Being at the centre of several identification procedures, the authorities considered these registers to be crucial civil documents. Reforms aimed at regulating parish registration encountered varying degrees of success in the two territories. However, regardless of whether they were successful or not, these initiatives were all inspired by the idea that religious registration was not reliable enough to guarantee the civil and legal value of the vital acts that were registered. The reforms were aimed at finding a more rational way of managing the parish registers, but without altering their essential nature as religious documents that performed crucial legal and civil functions in the field of personal identification.

Towards State Identities

The situation changed radically with the creation of pro-French governments in northern Italy after the Napoleonic military victories of 1796. Over the span of a few months, the new authorities installed in Milan set the foundations for a new system of personal identification and registration. The system was based on state identity documents and civil registers as the new tools by which citizens would identify themselves. The revolutionary idea of citizenship implied that the state needed to directly identify every individual under its jurisdiction, without any kind of mediation.[10] These fixed and certifiable identities were needed by the new civil authorities to determine who was part of the nation and who was not. According to the new order, parish registers and religious certificates were no longer appropriate for these civil tasks. Not only were they managed by an institution the revolutionaries looked upon with suspicion (the church). They were also seen as unreliable and defective sources of information. It this thus not surprising that the revolutionary authorities dealt with these issues in the same months in which they were building the foundations of the new state. Civil registers and identity documents were at the very heart of the new order that the Italian revolutionaries were creating.

The new identity documents were the first identification tools established by the pro-French authorities. In November 1796, only six months after the beginning of the French military occupation, the provisional administration in Lombardy created a residence certificate (*certificato* or *carta di residenza*). At a time when 'the nation takes up arms in defence of its freedom', the authorities needed a fail-safe way to identify those citizens who were opposed to the new government. As part of their attempts to quash potential opposition, they moved to expropriate the properties of all those citizens who had left the occupied territories. Lombard citizens who wanted to avoid having their property seized were given 30 days to go to their local municipality with two witnesses to request a residence certificate. These certificates would be valid for a period of three months and would be needed for the enjoyment of civil and political rights. The original aim of the certificates was not to identify their owners in their day-to-day lives but to guard them and – more specifically–their properties against political suspicion.[11] In this way, the law established a first meaningful connection between citizenship, identification, and property rights. In the following years, this residence certificate became a compulsory requisite for obtaining access to an increasing number of public functions: it was needed in order to be a boatman on the three major lakes of Lombardy, but also to vote for National Guard officers, or to enrol in the newly established military schools.[12]

A few days after the creation of these residence certificates, the municipality of Milan compelled every foreigner living in the city to obtain a security card (*carta di sicurezza*) from the local police. Unlike the residence certificates, security cards were mandatory not only in theory, but also in practice. The police decree specified that any foreigner who did not present their certificate upon inspection would be arrested as suspect, 'especially during the night'. Even if these original security cards did not include any physical description of their owners, their surveillance functions were quite clear: their aim was to reinforce negative discrimination against those individuals who were labelled as "foreigners" in Milan.[13] This form of discrimination had itself been present ever since the start of the French occupation. A few months before the decree on security cards was drawn up, for instance, general Despinoy had ordered the arrest of all those foreigners who could not justify their presence in the city and called for a special passport to be created for those not considered as suspect.[14] However, not all foreigners were equal. The French inhabitants of Milan – mostly employees of the occupying army – were required to carry a residence certificate instead of a security card. These certificates were nevertheless regulated directly by the French military command, who annulled their validity on a number of occasions in the following years.[15] In this way, the French inhabitants of Milan enjoyed a special condition, more similar to that of Cisalpine citizens. The French were still foreigners in the Cisalpine Republic, but with a special status guaranteed by the strong relationship between the state and its French counterpart.

In July 1797, less than a month after its creation, the Executive Directory of the Cisalpine Republic passed the first fundamental law on civil registration.[16] The decree was issued in a constituent phase of the republic. The two branches of parliament had yet to take office and the Directory itself was missing one of the five members prescribed by the constitution. This background is revelatory of the role that the law's proponents had in mind for civil registration. More than any other measure, civil registration was considered essential for the establishment of a new political order based on citizenship, rather than on birth right. As one member of the Cisalpine Parliament put it, a few months later, 'the civil and political status of each citizen is part of their property rights, actually it is the citizens' most precious property'.[17] Therefore, the civil status of citizens had to be regulated quickly.

This urgency to set up a system of civil registration is clearly visible in a letter sent by the committee charged with drafting the Cisalpine Constitution in the summer of 1797. The five members of the committee pointed out that the French Constitution, which was 'to be adapted to Lombardy', mentioned civil registers. Therefore, civil registers would have to be established to certify marriages, births, deaths, and citizenship status in every municipality or 'even better, in every parish'. The ultimate aim of these registers – in the committee's view – was for them to become the sole valid certificates of such vital acts 'in civil society', implicitly taking the place of parish registers.[18] Nonetheless, in the end, the Cisalpine Constitution only briefly mentioned civil registration, meaning that the legislative body was left to deal with the matter of its wider implications. However, the constitution made active citizenship possible only for those who voluntarily registered themselves at the municipal registers of citizens (*registro civico*).[19] Despite its theoretical goal to replace parish registers as sources of identification, the Cisalpine law on civil registration was prudent in this regard. Indeed, it did not contain any reference to the management of church registers, while it maintained the validity of other sources (such as parish registers) as legal proof of births, marriages, and deaths.[20] One year later, the interior minister's call to nationalise and transfer all the parish registers to the municipalities remained unheeded.[21]

By mid-1797, the Cisalpine government had laid the foundations of a new identification system. However, in the following years, the authorities struggled to effectively implement this system. The new republic's still not clearly defined administrative structures made the civil registers hard to be put in place. For instance, the rural territory around the city walls of Milan was not administratively organised until a specific law was approved in December 1798. The law was approved via an emergency procedure precisely because, without an administrative division of the territory, it would have been impossible to set up the new civil registers there.[22] These delaying factors, coupled with several legislative gaps, led to the civil registration system being still largely unimplemented even one year after its inception. During the summer of 1798, the interior minister acknowledged that the registers were being drawn up differently in the different departments – and often not according to the terms laid out by the law.[23] In May 1798, Francesco Reina, a member of parliament, pointed out that: 'the registers are not ready. [Civil registration] may be easy in the future, but not today'.[24]

Debates in the Cisalpine parliaments show how the new system of identity documents was seen as a priority by legislators, even when it was still largely unknown in the republic's territory. In November 1797, a member of the Great Council – whose name is not mentioned in the minutes – proposed to make the presence of two witnesses mandatory for the issue of a security card. At that point, another member curtly responded that this rule already existed and urged the anonymous representative to read the regulations before speaking.[25] A few months later, during a debate on the new conscription law, another councillor rhetorically asked his colleagues the following:

> how many will have [a residence certificate], considering that they do not have a permanent residence in any particular place in the republic, or that they still have the first security cards with them?[26]

Even if the members of the Cisalpine parliament were not well acquainted with the details of the legislation on identification, they were nevertheless quite aware of its inefficiency.

The uncertainties and continuous changes in the regulations made the implementation of the new identification system even more difficult. The 1797 law had created an incomplete system of civil registration, leaving several issues unanswered. In November 1797, the minister of the Interior pointed out some difficulties in the implementation of the registration system due to a lack of clarity in the law's language.[27] Moreover, the law could not enter into full effect until the approval of comprehensive legislation on marriage. The parliamentary debates over this matter mirror the Cisalpine élites' view of the civil registers as a fundamental alternative to those maintained by the church. In February 1798, the member of parliament and former priest Felice Latuada proposed that all marriages not registered in the 'municipal books' be declared void.[28] A few weeks later, his colleague Vincenzo Dandolo – of Jewish origin – submitted a bill on religious matters to the assembly, the second part of which was dedicated to the 'independence of the civil status of every citizen from the quality [type] of religion they embrace'.[29]

Even if these two proposals did not succeed, the marriage bill submitted by the legislative commission in May 1798 confirmed that no marriage would be valid unless it was celebrated with witnesses and registrars. Civil marriage registers were to be kept in the municipal archives.[30] A punishment of three years' forced labour was envisaged for those who tried to 'usurp the civil functions of marriage by trying to renew the same marriage for any other purpose or with any formula'. As it became clear in the subsequent discussion of the text, this provision was targeted at priests. The lawyer Luigi Oliva explained that these 'fanatics' were to be prevented from 'using the absence of some religious ceremonies' as a pretext to re-celebrate a marriage, since marriage was to be considered 'a mere civil contract throughout the whole republic'.[31] This bill, however, was also unable to secure parliamentary approval. The marriage law that was finally approved in the summer of 1798 did not mention civil registers at all. The first draft of the law did envisage a strengthening of the civil nature of marriage. One section of the draft mentioned civil registration as a prerequisite for any religious marriage and imposed a penalty of three years' forced labour on officiating priests who did not comply. Moreover, the draft law contained a clause reaffirming that 'from now on the only proof of marriage is the act registered at the municipality'.[32] However, the Great Council decided to remove these articles during the debate.[33]

In the same weeks, the Cisalpine parliament discussed new regulations on municipal administration. The bill presented by the former abbot Alvise Savonarola would have required each municipality to issue every one of the certificates related to the civil status of its citizens. In the debate, Latuada took the opportunity to reaffirm his secularist position on the subject

> We must not leave the books of registers [sic] of marriages and of other matters related to the civil status of citizens in the hands of the clergy. What relationship must the ministers of religion have with the regulations that interest society the most? The municipal administrations must be accountable [for civil registration] to the government and the nation.[34]

Despite the perceived urgency of the matter, the Cisalpine parliament could not overcome the technical difficulties related to the actual drawing up of the new regulations.[35] However, the secularist stance eventually succeeded when the executive directory decided to override the parliament and issued a regulation on civil registration. Several complaints had reached the government, lamenting 'influences external to the nature of the contract' (namely, the religious ceremonies), along with the complexity of the procedures. The government's decree, among other things, refused to recognise parish certificates as legal proof of the civil status of the bride and groom.[36]

In many other cases, the authorities opted for a pragmatic approach that allowed for the use of religious registers, even if civil registers were preferred in those places where they had already been set up. In August 1796, parish registers were used to identify those who had not yet been recruited for the newly established National Guard.[37] Two years later, civil registrars were also tasked with compiling a list of all men who had recently turned seventeen in order to facilitate their recruitment for the National Guard.[38] Both civil and religious registers were mentioned in the instructions for the compilation of conscription lists in November 1798.[39] In this manner, the government acknowledged the civil registers' inability to help determine which people were already recruitable. The importance of civil and religious registers for conscription purposes also led the Cisalpine councillors to pass a law in which a specific article was dedicated to penalties for 'those who will alter the registers of births, marriages, and deaths'.[40] Significantly, these penalties were not just targeted at those who tampered with civil registers, but also aimed to protect the religious registers. The Cisalpine politicians had to acknowledge the public function of religious registers – even if the clergy were allowed to keep managing these. Three years after its inception, civil registration remained a largely theoretical political project in the Cisalpine Republic.

The system of identity documents underwent a similar process of redefinition after the first decrees of November 1796. At the beginning of the following year, the municipality of Milan extended the formal obligation to obtain a residence certificate to all its citizens. A proclamation stated that having a residence card had become even more important 'for good public order, and for the security and tranquillity of every citizen'. In particular, the presence of a large number of prisoners of war in the city meant that it was 'extremely necessary [to have the possibility] to recognise those who are citizens with certainty, and to distinguish them from those who are not'. As had already happened with the security card for foreigners, all citizens had to always carry the residence card on their person. If they failed to do so, they would be arrested and considered to be suspect.[41] The proclamation's text already clearly spelled out the semantic shift which had taken place regarding the nature of the certificate. Originally intended as a device to guarantee and protect citizens and their property, the residence certificate had become a formal tool for the identification of individuals. Its function therefore became more similar to that of the security card, even if the presence of two different types of certificates with the same features implied – at least theoretically – different levels of respect and treatment for citizens and foreigners.

In January 1798, the Cisalpine government made another relevant step toward establishing an equivalence between the identities of citizens and foreigners. An important proclamation made by the new minister of Police, Fedele Sopransi, included new rules both for residence certificates and security cards. Here again, the stated cause of the provision was the extraordinary number of vagabonds and foreigners who were threatening public order. To locate these 'disruptors of every social order' while safeguarding 'honest travellers', the minister introduced a series of innovations. The legislation on security cards was extended to the whole national territory, thus losing its original, urban,

"Milanese" feature. At the same time, having a residence certificate became a requirement for every citizen willing to travel within the republic, '[because] Cisalpine [citizens] must also all have a document that justifies their qualities'. Citizens without identity documents were to be immediately arrested as 'suspect people'.[42] The innovations introduced by this proclamation effectively equated the legal standing of citizens and foreigners. Even if they had to use different documents to certify their identity, both categories were subject to the same obligations for travel throughout the republic. The difference between their rights and duties had lasted just over a year. Eventually, citizens ended up being treated in much the same way as foreigners – which implied a substantial loss of rights for them.

In the summer of 1798, the new minister of Police Vincenzo Brunetti confirmed this stance in a proclamation that reorganised the different identification devices into a more consistent whole. According to the minister, society could effectively be divided in two parts. On the one hand, there were those individuals who were faithful to the republican values of 'peaceful labour, the attachment to true democracy, the exercise of social duties'. On the other, there was an increasing number of 'vagabonds, idlers, adventurers' who were

> the effective means used by our enemies to spy on and slander the performance of the government, to foster discontent among citizens and discord between friendly people; they are always the partners of thieves, the disturbers of public tranquillity, the corrupters of morals.

Therefore, Brunetti thought it necessary for each person travelling within the republic to justify their presence there and their behaviour.[43] In his proclamation, there is a complete overlap between the political acceptance of the republican order and a display of productive behaviour. Being a good citizen meant being a productive citizen. The identification system was therefore crucial in the definition of the boundaries separating good from bad citizens and – more broadly – individuals from each other, within the new political order.

After a year of Habsburg occupation in 1799–1800, the reinstated Cisalpine authorities continued to favour the evolution which had already begun during the republic's first phase of existence. The distinction between the rights and duties of foreigners and citizens became even more subtle, while the nature of the identification system almost completely lost its original civil features. In June 1800, the departmental administration of the Olona made security cards mandatory for both citizens and foreigners in the department's capital, the only difference for these two categories being the issuing authority.[44] A few days later, the police of the same department – provisionally responsible for public order throughout the republic – extended this obligation to the entirety of the national territory, thus turning security cards into the fundamental means of identification for every Cisalpine citizen. Quite interestingly, the proclamation presented this provision as a way of guaranteeing the republic's citizens against 'every arrest or other inconvenience [caused] by [police] patrols'.[45] The original concept underlying the identification system had therefore been completely overturned. When the Cisalpine Republic was first instated, citizens needed identity documents to prove their rights, whereas, by the start of the Second republic, they needed these documents to avoid police suspicion.

After the founding of the Italian Republic in 1802, the new minister of the Interior issued a set of rules for the police which completed the previous trajectory of the identification system and laid the ground for its further evolution. The security card was confirmed as the republic's main identity document. According to the new rules, every citizen or foreigner residing in the republic who wanted to travel needed to obtain a card 'to avoid any doubt that may arise regarding their identity'.[46] The process by which citizens and foreigners were placed on an increasingly equal footing was now finally achieved. The security card replaced the original Cisalpine system, based on two different identification devices. There thus was a coherent thread, weaving its way through the chaotic and longstanding processes of reform and arriving at these means of identification. Born as identification devices with two separate aims – guaranteeing citizens on the one hand and monitoring suspects on the other – the residence and security cards slowly converged toward a single goal: to materialise the distinction between "good" and "suspect" individuals.

The administrative laboratory of the Cisalpine Republics reveals the culture of identification that the post-revolutionary authorities envisaged. Their new form of citizenship implied a certain distinction between who was part of the nation and who was not. Consequently, they needed to develop identification tools and procedures that could establish fixed and certifiable identities. This process could not be delegated to the clergy, who had traditionally managed registers and identity certificates. The Cisalpine politicians were keen to specify that religious sacraments had no civil and legal value. Moreover, the new identity documents catered to the need not only to certify individual identities, but also to evaluate their reliability. Even if these reforms went largely unimplemented, they are a sign of the emergence of a mature culture of identification.

A Cumbersome Model

The Italian post-revolutionary and Napoleonic governments always looked to the French identification and registration systems as reference points. This was not a simple matter of political dependence or subalternity. France – the cradle of a new way of thinking of the state – acted as a model for the new political order that the Italian élites were trying to implement in their own states. Moreover, France played a significant role in guaranteeing the self-same existence of these states, which were under its military protection. The *Grande Nation* was also, at the turn of the century, among the European states with the most efficient administrative traditions.[47] This was especially the case as far as regarded issues of registration and identification. As we have seen, the Kingdom of France had already taken control of local parish registers in the late seventeenth century, progressively increasing its authority in the field over the following hundred years. The same can be said for identity documents, at least from the beginning of the eighteenth century. In the countryside, royal authority had slowly "incorporated" the local agents that oversaw identification procedures: the parish priests and the police force known as the *Maréchaussée*. Moreover, in subsequent decades, the French government had developed new identification technologies and tools, such as the use of physical descriptions on passports and pre-printed discharge papers.[48] The French state was thus at the forefront of technologies and procedures of identification.

The Italian élites, however, also worked on adapting the French model, hybridising it with their own administrative culture and adjusting it to what they saw as their own, different context. If the Italian governments' culture of identification owed much to the French example, it is also true that it was the result of a permanent process of redefinition based on autonomous political considerations and administrative traditions. The French governments allowed their satellites a large degree of freedom in the development of their own forms of state building. In the Cisalpine and Italian republics, local élites took advantage of this freedom and elaborated administrative solutions that modified the post-revolutionary model. However, as imperial control over the Italian governments became more and more strict over time, the agency of these "native" élites began to lose its impact.

This was true, first of all, for civil registration. The 1797 Cisalpine law was clearly inspired by its French counterpart, which had been passed by the Legislative Assembly in September 1792.[49] Of the 47 articles making up the Cisalpine law, 36 were adaptations of articles from the French law, with minor changes. Nevertheless, the structure of the law itself demonstrates ways in which the Milanese authorities adapted the French model to the Italian context. In particular, the Cisalpine law left civil marriage substantially unregulated and allowed parish registers to act as legal sources. These differences show how the government considered the relationship between the new registration system and religion to be a delicate matter: a word-for-word transcription of the French law might have offended the religious sensibility of large sectors of Italian society.

Moreover, the Cisalpine law also differed from the French one on some technicalities. In France, the registrar wrote the same acts on two different registers, the second of which was sent to the prefecture. The Cisalpine registers, on the contrary, were to be compiled on a register in which each sheet of paper contained two identical forms, one of which was meant to be given to the involved parties as a certificate.[50] A few Cisalpine and Italian politicians even dared to criticise the French model. As part of his proposals to reform the Cisalpine system, the minister of Worship Giovanni Bovara pointed out some deficiencies in the French model of civil registration. In 1802, for instance, he acknowledged that the French civil marriages often lacked basic decency, emphasising the usually poor appearance of the registrars and the city halls.[51] On another occasion, he highlighted that the Italian context was different from the French one, and that the government was therefore well aware that any imitation of the French legislation would be a 'bad imitation'.[52] Even if France acted as a model for the Cisalpine system of civil registration, Italian politicians could also display a certain lack of deference to the French administrative tradition.

This autonomous attitude changed in 1806 when the French Civil Code – that also regulated civil registration – was applied throughout the Kingdom of Italy. At first, politicians in Milan tried to maintain the practice of compiling the registers in two columns, following the Cisalpine tradition, but they eventually aligned with the French model.[53] In the following years, the Italian system of civil registration became ever more integrated with the imperial one. In 1812, for instance, the minister of Justice Giuseppe Luosi proposed a decree that contained several disciplinary provisions that had been adopted in France over the previous years. The decree stated that:

> considering that the Italian legislation on civil registration is shared with the French one, it is advisable to adopt and observe all French disciplinary provisions also in the [Italian] kingdom.[54]

The Milanese politicians mobilised French authority on this matter for other purposes too. Before he officially proposed to pre-print civil registers in 1807, the minister of the Interior Ludovico di Breme asked the foreign minister if this practice had also been adopted in France. At the same time, the minister of Justice, Luosi, took care to mention French legislation when opposing the self-same proposal.[55] Some months later, the State Council's committee for the management of municipal administration put forward a proposal to charge citizens who requested excerpts from the civil registers. The practical aim of the proposal was to replenish municipal coffers, but to give it strength the commissioners highlighted that the proposed fee was already common practice in the empire.[56] Italian authorities relied on the French model not only to support their proposals, but also to solve their doubts. When unsure about how to answer a question from the prefect of the Rubicone – regarding who was to choose the officials in charge of the civil registers in larger municipalities – the minister of the Interior Luigi Vaccari asked his French counterpart for advice.[57] After the introduction of the Civil Code, French legislation thus went from being a simple source of inspiration to becoming the bedrock on which the Italian system of civil registration was built.

The system of identity documents underwent a similar process of slow assimilation. The cards established in the Cisalpine Republic in November 1796 were only loosely based on their French counterparts. On the one hand, the name given to the Cisalpine security cards revealed their French inspiration. In September 1792, the French National Assembly had instituted a certificate aimed at identifying each one of the male inhabitants of Paris. After some months, it became known as the *carte de sûreté*. During the following year, every man in Paris, in theory, received a security card. According to the law, failure to carry this device on one's person at all times could lead to six months in prison.[58] On the other hand, while the Cisalpine security cards were, like their French counterparts, only used in the republic's capital, they differed from the latter in that they were initially prescribed only for the foreign residents.[59] Moreover, the existence of a separate identity document for citizens – the residence card – had no equivalent in French legislation. The differences became even more striking when the security cards became mandatory for foreigners throughout the republic's territory in January 1797.[60] In the French republic, the only national identity document was the internal passport established in 1792.[61] However, its features differed in several aspects from those of the Cisalpine security cards. During this first phase then, the Milanese authorities created an identification system that was inspired by, but not modelled on, the French one.

In the following years, the Cisalpine identification system strengthened its particular features, even if Milanese politicians occasionally proposed changes that would make it more like the French one. In late 1797, the democrat and former general Giuseppe La Hoz suggested adopting the French model based on internal passports for foreigners, but the Great Council ignored his proposal.[62] Some months later, the Cisalpine councillors discussed a bill that would have extended the passport system to all travellers, including citizens, also following the French model. The bill clearly described the appearance and shape of the proposed new passports, in order to make them uniform across the republic. Its provisions were intended to remain valid 'until the [arrival of] general peace'.[63] During the debate on the bill in the Great Council, several representatives proposed to make citizens exempt from obtaining these new passports since they already had residence certificates. These representatives argued that the law's effects could have been attained simply by making the layout of the residence certificates uniform across the republic. Divided on this regard, the councillors agreed that the passports would all have to be signed by the issuing authority, just as they were in France, 'to avoid doubts and forgeries'.[64] This second attempt to bring the Cisalpine identification system in line with the French one failed when the Council of Seniors rejected it after a lengthy debate.[65]

However, a few weeks later, the new minister of Police Vincenzo Brunetti issued a decree that included some of the suggestions from the previous debate. The announcement stated that Cisalpine citizens were now required to carry their residence certificates with them when travelling within the republic's territory, with immediate effect.[66] In this way, the residence cards became more similar to the French internal passports, without however losing their own specificity. The 1802 police reform confirmed this trend. The new security cards became a necessary tool for travel throughout the republic's territory both for citizens and for foreign residents, just like the internal passports in France. At the same time, these cards also functioned as identity documents for the sedentary population, like the *cartes de sûreté* in larger

French cities.[67] The Cisalpine identification system turned the security cards into theoretically universal certificates, while, in France, a functional distinction was maintained between internal passports and security cards.

Just like the civil registration system, the identification system also became more closely aligned with the one used in France after the monarchical turn. In May 1809, the police director Diego Guicciardi changed the forms on which Italian passports were compiled to make them just like the French ones, by adding a field indicating how many days the owner had been given to exit the national territory. The stated goal of the change was to 'establish a method similar to the one that is practised in the French empire'.[68] The real turning point, however, was the 1811 reform that abolished security cards and replaced them with a dual system explicitly based on the French model. In the report accompanying the projected reform, the new police director Francesco Mosca explained that the extension of the French Penal Code to the Kingdom of Italy had made this reform urgent. All of the project's provisions on passports were openly inspired by French rules. The only substantial difference was the looser degree of control the project envisaged for foreign travellers who were just passing through the kingdom. These were far more common than in the French empire because of the Kingdom of Italy's geographical location. Moreover, imperial citizens enjoyed some privileges, such as exemption from obtaining visas that would have 'proven too much of an impediment for the multiple relationships that our kingdom must maintain daily with them'. Mosca's project also introduced internal passports as documents needed for travel outside one's department of residence, following the French model.

However, the reform also created a type of document that had no equivalent in the French imperial system: the registration certificate (*certificato di iscrizione*). This document was mandatory even for those who never left their departments and only vaguely resembled the *cartes de sûreté* that were used in some of the larger French cities. In particular, the certificate was strictly connected to a new tool for the civil registration of residents, the general population list (*ruolo generale di popolazione*).[69] Though not formally implemented yet, this was introduced in 1809 as a fiscal tool. The general population lists were to be drawn up and periodically updated by local municipalities.[70] This peculiar feature of the lists provided a link between civil registration and identity documents, perpetuating the longstanding original administrative tradition of the Cisalpine and Italian élites. The Legislative Council itself, when called to discuss the 1811 reform, invoked Italian autonomy on the matter by expressing the opinion that the 1802 system was enough to guarantee public order.[71] However, the reform was eventually signed into law directly by the viceroy as a decree. This was a firm step forward uniformity with the imperial legislation.[72] The following year, police director Giacomo Luini moved to allow French imperial citizens to enter Italian territory with just a French internal passport.[73] The Napoleonic system was becoming increasingly integrated, and the culture of identification promoted by the Italian élites was inevitably forced to converge with the French model, even if it never completely lost its own specificities.

A Troublesome Implementation

If the Italian post-revolutionary élites shared the aim of establishing certifiable and fixed identities, they were much more divided on the subject of how this goal was to be achieved in practice. A recurring matter of friction was, for instance, how to combine the securitarian features of the identification system with the need to ensure freedom of trade and people's civil rights. One of the main causes of the frequent changes in the Cisalpine identification system was precisely this contradiction. In June 1798, for example, the Great Council discussed a bill on passports whose aim was to protect the republic from its enemies, while at the same time taking care not to 'vex and annoy the quiet citizen or the useful merchant'.[74] The same preoccupation led the government to establish a temporary passport (*carta di passo*) for those seasonal workers who often had to cross the border for work.[75] Over ten years later, when faced with the fact that document checks were carried out with excessive zeal, police director Giacomo Luini made clear that the aim of the identification system was to 'detect wicked and suspect individuals, not to bother honest and good citizens'.[76] Despite regime changes and shifts in the political climate, the contradiction between the control of identities and freedom of movement remained relevant for these Napoleonic politicians.

However, the implementation of the state culture of identification was especially difficult in the field of civil registration, given the latter's relationship to the clergy. As we have seen, the Cisalpine authorities agreed on the fact that parish priests could not be left in charge of civil registration. Not only did they consider the clergy to be politically suspicious, they also wanted to establish a precise distinction between vital acts with legal consequences – such as births, marriages, deaths – and the religious ceremonies that were connected to them. Moreover, parish priests had largely proved to be unreliable registrars in the past. Thus, the political project of establishing a system of civil registration remained a priority for the Cisalpine authorities. However, as the years went by, it became increasingly clear that the state did not have the means to achieve this ambitious political project.

The government acknowledged the inadequacy of the new system of civil registration in the years after the brief Habsburg occupation in 1799–1800. The authorities gave up on their original aim of diminishing the legal value of religious registers, and began to explore ways to use them as civil sources. In September 1801, a few months after the re-establishment of civil registration, the government decided to collect a copy of every parish register once a month. The goal of the practice was to 'provide important and detailed knowledge of births, deaths, and marriages to the government' – an implicit admission that the civil registration system was not working as planned.[77] This obligation to send copies of the registers to the government every month put a substantial strain on parish coffers, so much so that the minister of Finance Ambrogio Soldini had to intervene to make the postage of these registers free of charge – in this way, formally equating the parish registers to public documents.[78] In the same period, the authorities also chose parish registers and certificates as their main sources for the compilation of conscription lists.[79] In 1805, moreover, the government made it mandatory to use officially stamped paper for parish certificates, thus strengthening their official and legitimate nature.[80] On a practical level, the original secularist stance had already essentially been superseded: the parish registers and certificates had gained a recognised civil function.

However, this pragmatic tendency continued to find opposition within the Italian legislative bodies. The acknowledgement of Catholicism as the state religion in the 1802 constitution of the Italian Republic and the subsequent signature of the Concordat between the Republic and the Holy See in 1803 weakened, but did not eliminate, the secularist leanings of part of the Italian élite. For six years, from 1802 to 1808, the members of the national government debated the possibility of involving the parish priests in the civil registration system. These debates, as we shall see, are quite revealing of the differences of opinion that divided Italian politicians when it came to implementing their shared state culture of identification.

A few months after the establishment of the Italian Republic, the newly appointed minister of Worship Giovanni Bovara submitted a general plan for the organisation of civil marriages. The 68-year-old abbot was no stranger to political engagement. He had been one of the most influential officials in Habsburg Lombardy in the field of religious affairs. As minister of Worship in the pro-French governments, he embodied a prudent reformist stance, combining the principles derived from the Josephinism of his youth with the new Napoleonic guidelines.[81] Bovara's project for civil marriage contained, among other things, a proposal that would have been hardly conceivable a few years earlier: the possibility of delegating the management of the civil registers to ministers of religion, be they Catholic or not. Bovara was convinced that the civil registers were a crucial tool for explaining the difference between civil contracts and the sacraments to the general population: as he wrote in his report, 'it is necessary to rectify the practices in order to rectify the ideas'. However, the separation between religious and civil marriage had to be effected in the easiest way possible, without going against the habits of the people themselves. The involvement of religious ministers would have made it easier for the general population to accept the civil registers, while at the same time providing an antidote to the endemic lack of bureaucratic personnel in the countryside. The minister stated that municipal officials were unqualified for carrying out civil registration not only because of their illiteracy, but also as a consequence of their lack of decorum. "Sloppy" civil procedures had to be avoided by all means: it was necessary to make civil marriage as similar as possible to a religious ceremony by giving it more 'solemnity'.

Nevertheless, the bill also envisaged the possibility for citizens to avoid having to recur to a minister of religion for civil registration, as it allowed them to also register their vital acts before the municipal registrar.[82] In this way, the new law would have set up a double system of civil registration in each municipality, with two parallel sets of civil registers (one managed by the civil registrar and the other by the parish priest). This solution was probably envisaged to cater to the secularist stances still held by many in the Italian élite. At the same time, Bovara's bill was not just an effort to compensate for the deficiencies of state bureaucracy. It also had the aim of using priests as a tool by which to make civil registration a more solemn act, therefore increasing its acceptability among the general population. In this way, the proposal outlined a sort of symbolic transfer of authority from the clergy to the bureaucracy – and, more generally, from the church to the state. The result of these contradictory goals was Bovara's proposal for a civil registration system organised in a semi-communal manner. This would have probably overcome the weaknesses of the bureaucracy, but only by abandoning the Cisalpine élites' original idea of creating a registration system exclusively managed by state agents.

In the same months, however, one of Bovara's fellow ministers was also elaborating a proposal regarding Cisalpine civil registration, and a much more radical one at that. The minister of Justice Giovanni Bonaventura Spannocchi, a jurist with strong links to the pre-revolutionary juridical tradition, drew up a draft Civil Code that envisaged giving civil legitimacy to religious marriages. According to his proposals, parish registers would fulfil all the functions of civil registration.[83] This would have implied a clear step backwards in the separation of civil contracts and sacraments. It is not surprising, then, that the minister of Worship was in a hurry to debate his own bill, though this was eventually blocked by the imminent signature of the Concordat.[84] As Bovara pointed out in a long report on the bill, his project was to combine the 'healthy principle of jurisprudence' with respect for the opinions and habits of the general population. Since Catholicism was the state religion, it was necessary for this religion to have a certain influence on the basic principles of civil legislation. This influence, however, was not to be exerted by handing over the control of civil rights to the religious authorities, but rather by taking into account the opinions and customs of the general population. He was aware that this compromise could disappoint 'a rigid philosopher'. However, Bovara was sure that his solution would not have affected the civil jurisdiction in any way.[85]

Bovara's bill finally came up for debate in the Legislative Council in March of 1803. The Council's judiciary committee, composed by Tommaso Gallino and Ignazio Magnani, presented an extended report on the proposal, ending with a negative opinion of it. The councillors considered it unwise to delegate civil registration to the parish priests, because this would have made them the only arbiters in the definition of marriage rights. If the law had been approved, priests would have had the power to refuse marriages in cases that were allowed by the civil registration, and to celebrate ones that were not allowed by the republic's laws. In order to solve this issue, the committee proposed to make it mandatory for a priest to be present whenever two Catholics contracted a civil marriage.[86] The subsequent debate in the Council did not consider this last proposal. The councillors only discussed the overall principles of the bill before postponing the discussion of the articles themselves. Aldo Aldini noted that Bovara's bill would have given the Catholic faith precedence over civil law, while Carlo Testi stated that a civil law on marriage should not have tried to regulate its religious aspects. All the councillors agreed to not having the validity of civil marriages depend on priests.[87]

Bovara's proposal thus encountered a new front of opposition in the Council. The debate highlighted the persistence of secularist stances among the councillors, while the report from the judicial committee raised more technical questions. The proposal put forward by the minister of Justice, on the other hand, had been far more radical in its involvement of the clergy. Thus, Bovara's moderate proposal found itself caught in the crossfire between different conceptions of civil registration. Regardless of whether Italian politicians recognised the problems inherent in the Cisalpine Republic's system of civil registration, they disagreed even more profoundly on possible solutions.

The approval of the Concordat in November 1803 prompted Bovara to reclaim the initiative. A few months later, he presented a slightly different bill, still based on a "communal" solution.[88] Once again, a lively debate arose over the role that priests were to play in civil registration. The councillor Ambrogio Birago proposed to specify that civil marriages had to be registered in a municipal office, in order to prevent the priests who officiated them from doing so in church. Moreover, Birago appealed to the French law to argue against the creation of a second set of civil registers in the hands of the priests. Birago's secularist stance found support in the Council, even if the article under discussion was eventually approved with no amendments.[89] Thus, Bovara's attempt to mediate between the secularisation of the registers and "religious" public opinion still found opposition in the Council. However, this time, no one questioned the overall idea of involving priests in the registration process. Similarly, no voices were raised against the preferential treatment of Catholicism compared with other religions. The approval of the Concordat had probably weakened secularist positions in the Legislative Council.

Faced once again with failure when this slightly amended bill was rejected, Bovara drafted a new proposal which contained substantial changes compared with his previous ones. Quite significantly, he presented this third attempt at reform in the form of a vice-presidential decree. Bovara thus hoped to bypass the Legislative Council – that had voted down his previous proposals – by convincing the vice-president Melzi to sign the law into effect on his own authority. This latest bill called for the civil registers of marriage in small municipalities to be delegated to local priests. Unlike in the previous drafts, a municipal registrar was not envisaged – thereby implying that parish priests were to be the only registrars in the republic's villages. While the previous reforms were aimed at creating a double system of civil registration (one managed by the clergy, the other by municipal authorities), this one adopted a different perspective. The process of registration was to be split between civil registrars and parish priests along practical lines, depending on the size of each municipality. Bovara had therefore given up on finding ways to use the clergy to turn civil registration into a more 'solemn' affair. This proposal only focused on finding efficient registrars in places in which the state could not count on the efficiency of its bureaucracy, mainly in the countryside.[90] However, despite positive opinions from the minister of Justice and the minister of the Interior, this third attempt at reform was also struck down.[91]

This marked the end of the attempts to reform the civil registration system inherited from the Cisalpine Republics. A few months later, the French Civil Code was adopted by the Italian Republic, changing the legislative framework within which civil registration operated. During the previous years, secularist stances among the Cisalpine/Italian élites had gradually weakened. It was now clear enough that civil registration could not be implemented without somehow involving the local parish priests. Already in May 1805, the possibility of involving parish priests in civil registration was raised by the viceroy Eugene himself in the State Council. He noted that, in France, the registers had been assigned to civil authorities for the only reason that 'five million heterodox people' were living in the country. On the contrary, almost everyone in the Kingdom of Italy was baptised and buried with Catholic rites. The only problem, in the viceroy's opinion, was that the church had its own specific legislation on marriage and '[did] not want to bend to civil laws'. Therefore, he suggested that marriage registers be assigned to the municipalities, while the civil records of death and birth could be delegated to the clergy.[92] This proposal was consistent with the tendencies Italian politicians had displayed in previous years. Moreover, if approved, it would have set a precedent that would further increase support for involving the clergy in civil registration procedures. Starting with Eugene's proposal, that was ultimately not implemented, the idea that civil registration was needed as a way to respect religious minorities was sidelined.

The need to implement the French Civil Code in Italy made the discussion even more urgent. As a matter of fact, the Code itself did not give any specific instructions as to who was to act as registrar.[93] During the preliminary discussions between ministers on how to implement the Code, Bovara made it clear that he was now against any compulsory involvement of the clergy in civil registration. Priests, he argued, could not be registrars if this position implied that they would be forced to register divorces or those civil marriages that were not allowed by canon law.[94] Bovara – who had proposed to involve the clergy in civil registration procedures three times in the previous three years – had radically changed his mind after the approval of the Code Napoleon. Over the following years, he was to be the harshest opponent of any kind of innovation in this direction.

The Legislative Council started to debate the draft rules for implementing the Civil Code in mid-March 1806, only two weeks before the new Code was to be activated throughout the kingdom. Not surprisingly, several councillors raised the possibility of delegating the civil registers to priests in small municipalities, to bypass the inability of local officials there. Stefano De Bernardi – who was leading the discussion – strongly opposed this proposal. If it was approved, he argued, it would become impossible to eradicate the confusion between civil acts and religious ceremonies from popular mentality. This secularist stance was shared by Luigi Lambertenghi, who, nonetheless, pointed out that the mayors of most small municipalities were practically incapable of carrying out registration procedures. The Council thus modified the original draft, allowing for the designation of an adjunct registrar in these places, as long as the latter was a municipal official – thus ruling out any involvement of the clergy in civil registration.[95]

The following day, the Legislative Council met in the presence of the new minister of Justice, Giuseppe Luosi. The debate soon returned to the question of who was to act as registrar in the smaller municipalities, where even De Bernardi admitted that it would have been difficult to find proper administrative personnel. The councillor Carlo Verri – who had not intervened in the previous day's debate – once again raised the possibility of involving priests in the registration procedures, pointing out that this had been done for several centuries. The minister of Justice shared the Council's concerns, but considered any further involvement of the clergy to be problematic, especially in matters regarding divorce. He nevertheless admitted that some priests might have no qualms in dissolving marriages – and that these particular priests could be involved in the procedures. At this point, the Council approved a new amendment to the original draft that allowed prefects to appoint any citizen who was domiciled in a given municipality as an adjunct registrar, including priests.[96] In the following session of the Council of State, discussions once again focused on the possibility of involving priests in the registration procedures. The minister of Justice Luosi acknowledged that delegating the management of the registers to priests was against the 'general principle' of the Code, but nevertheless supported the proposal.[97] A provision delegating the management of the registers was ultimately approved alongside the regulations for implementing the Civil Code.[98] The councillors' attitude shows how secularist stances had, by now, given way to the practicalities of finding a way to solve the conundrum of implementing the civil registration system in small municipalities. Civil registration would have been impossible without the clergy.

However, the minister of the Interior Ludovico di Breme was not satisfied by the new rules, as they only allowed cooperation from the clergy on a voluntary basis. Faced with increasing difficulties in implementing the civil registration system in smaller municipalities, di Breme proposed to make this cooperation mandatory when needed.[99] Bovara did not waste any time in expressing his opposition to this new proposal. The new legislation – which included the institution of divorce – made it impossible to force priests to accept a task to which 'they had [already] declared their opposition'. The forced co-optation of priests as registrars, he argued, would have created confusion between civil acts and religious sacraments.[100] Paradoxically, this had been one of Bovara's goals when proposing the inclusion of the clergy in the registration system a few years earlier. Di Breme was thus forced to withdraw his proposal and accept a more minimalist solution: he sent out a circular urging parish priests to be more active in their cooperation – though this was still voluntary.[101] Nevertheless, the interior minister did not resign himself to supporting the status quo. In December 1807, he took advantage of viceroy Eugene's presence in the capital to try to get him to sign a decree that would have bypassed Bovara's predictable opposition. This latest proposal called for the compilation of civil registers of births and deaths to be delegated to parish priests, while leaving the registers for marriage and those that dealt with non-Catholics in the hands of the municipalities. The minister adopted pragmatic arguments to sell the proposal. The registrars' 'ignorance' had made civil registration inefficient, especially in smaller municipalities. Moreover, many registrars had requested a salary, imposing a significant burden on municipal budgets. The involvement of priests would have solved these practical issues.[102]

The interior minister's attempt to bypass his colleagues, however, failed. In January 1808, the draft decree was submitted – probably by the viceroy himself – to the Legislative Council: the latter decided to request opinions from the ministers of justice and worship.[103] The minister of Justice, Luosi, opposed the proposal due to technical reasons. He believed that di Breme's bill would not have solved the problems related to the inefficiency of the municipal personnel but that it would have compromised the uniformity of the civil registers.[104] Bovara added considerations of a strictly political nature to this opinion opposing the proposal. He said it would have created a distinction between Catholics and non-Catholics, caused by an issue that was unrelated to the nature of civil registration: religion. This distinction might upset families, especially those who were grieving. Furthermore, the civil registers had been created precisely to eradicate the confusion between civil and religious jurisdictions that was rooted in public opinion. Delegating registration to priests would have meant a step backwards, against the spirit of the law. Once again, Bovara used secularist arguments to oppose any further involvement of the clergy.[105] Most of Luosi and Bovara's arguments were shared by De Bernardi's report to the Legislative Council. In his opinion, the draft decree contradicted the Civil Code itself. Moreover, its stated goal of saving money was less relevant than other political considerations. It was much more important 'to destroy, in the opinion of the people, the religious ideas that are attached to the aforementioned acts'. The Council eventually approved of these arguments and formally rejected di Breme's proposal.[106]

After the rejection of this last proposal, there were no further proposals to delegate civil registration procedures to the clergy during the Kingdom of

Italy's remaining years of existence. The satisfactory compromise reached for the 1806 regulations and Napoleon's break with the pope in 1809 probably discouraged any further proposals in this respect. However, the debates that had emerged among the ranks of the Italian political élite in the previous years are indicative of the tensions that existed within the state culture of identification. If the idea of creating certifiable and secularised civil registers – and with them certifiable and secularised personal identities – was shared by most Italian politicians, their views diverged greatly on how to put this idea into practice. Other political cleavages intervened in this matter: not only secularist and pragmatist stances, but also Bovara's will to avoid any further discontent among the clergy. When it came to its implementation, the state culture of identification proved to be a fertile ground for political debate and conflict.

*

Revolutionary and Napoleonic Italy witnessed the coming of age of a state culture of identification. The post-revolutionary élites needed certifiable and fixed identities in order to define who was part of the nation and who was not. In their view, the imposition of certifiable identities also carried substantial legal and economic benefits. These needs combined with the will to make society easier to read – and thus easier to govern. In developing these new tools, the civil authorities did not start with a blank slate. They could count on the example of the church's identification culture that had provided tools for the construction of certifiable identities for centuries – even if these tools were largely defective. However, the new civil élites wanted to manage the identities of their citizens/subjects directly. They needed new tools and procedures to emancipate themselves from the church. The creation of identity documents and civil registers was the logical consequence of their culture of identification.

The basic axioms of this culture remained stable throughout the Revolutionary and Napoleonic period. Each individual was to have a fixed identity, certified by the state via precise registration and identification procedures and tools. However, there were many different ways in which this shared culture of identification could be implemented. Changes in the political climate directly influenced how the Italian élites conceived of their reforms. Identity documents were first introduced as a way of certifying who was a citizen and who was not, but in a few short years, they became tools by which to separate society into reliable and unreliable individuals. Secularist stances on civil registration remained powerful in the years before the Concordat but went into a steady decline afterwards. Moreover, the sources of this culture of identification were different and – from time to time – even contradictory. The civil authorities saw the French model as an inspiration, but always thought it legitimate to adapt it to their own administrative traditions and to the Italian context. Even after the introduction of the French Codes, they continued to develop original solutions. Ultimately, the Italian state culture of identification was flexible enough to survive until the end of the Napoleonic governments.

Notes

1 *Mémoires de Jacques Casanova de Seingalt écrits par lui-même. Édition originale, la seule complète. Tome cinquième* (Paris: Ernest Flammarion éditeur, 1832), 108–110.
2 Law (18 December 1800, 27 frimaio IX), *Raccolta* 2/I 195–197. Ten years later, the Penal Code punished whoever used a passport with a fake name with up to one year's imprisonment: *Codice dei delitti e delle pene pel Regno d'Italia. Edizione ufficiale* (Milano: Dalla Reale Stamperia, 1810), 55.
3 Denis, *Une histoire de l'identité*, 335–337.
4 On Habsburg reformism, see Judson, *The Habsburg Empire*, 63–71.
5 Draft letter (9 May 1769), ASMi, Popolazione p.a., b. 16.
6 Draft circular to the Lombard bishops (30 October 1770), *ibid.*
7 Aulic decree (22 February 1782), summarised in *Collezione di I.R. leggi ed ordinanze in materie economiche ecclesiastiche dall'anno MDCCLXVII fino al MDCCLXXXIII tradotte dal tedesco. Parte prima* (In Pavia: Nella Stamperia del Regio ed Imperial Monistero di S. Salvatore, 1785), 133.
8 «Notizie del Mondo», n. 57 (16 July 1785) e n. 59 (23 July 1785).
9 Claudio Povolo, *Dimensioni del sacro e dimensioni del profano. I registri canonici tra istituzioni ecclesiastiche e secolari nella seconda metà del Settecento*, in *Archivi e cittadino. Genesi e sviluppo degli attuali sistemi di gestione degli archivi*, ed. Gianni Penzo Doria (Sottomarina, Il Leggio, 1999), 39–72.
10 Noiriel, "Surveiller les déplacements ou identifier les personnes?," 82–84.
11 Announcement by the general administration of Lombardy (11 November 1796, 21 brumaio V), in *Raccolta*, 1/II, 132–133.
12 Announcement by the general administration of Lombardy (27 November 1796, 7 glaciale V), in *ibid.*, 154–157; announcement by the municipality of Milan (1 March 1797, 11 ventoso V), in *ibid.*, 273; announcement by the minister of War (21 June 1798, 3 messidoro VI), in *Raccolta*, 1/V, 156.
13 'Provvidenze per la polizia della Città di Milano' (20 November 1796, 30 brumaio V), in *Raccolta*, 1/II, 147–148.
14 Announcement by general Despinoy (28 May 1796, 9 pratile IV), in *Raccolta*, 1/I.
15 This certificate – originally named *Biglietto d'alloggio* – was issued directly by the military command: 'Decreto del Comandante della Piazza' (25 April 1797, 6 fiorile V), in *ibid.*, 383–384.
16 'Legge relativa ai Registri' (24 July 1797, 6 termidoro V), in *Raccolta*, 1/III, 88–91.
17 Session of the Great Council (6 February 1798, 18 piovoso VI), *Redattore*, VI/32, 501–502.
18 Letter from the costitutional commitee to the president, ASMi, Popolazione p.a., b. 16. This letter is dated '21 fiorile' (10 May 1797), even if the various provisional committees were established only two months later (9 July 1797, 21 messidoro V): *Raccolta*, 1/III, 69–70.
19 Constitution (8 July 1797, 11 messidoro V), *Raccolta* 1/IV, I–XX. The new constitution imposed by the French ambassador in September 1798 once again mentioned civil registration only in connection with citizenship: Constitution (1 September 1798, 15 fruttidoro VI), *Raccolta* 1/V, 280–297.
20 'Legge relativa ai Registri' (24 July 1797, 6 termidoro V), in *Raccolta*, 1/III, 88–91.
21 Report from the Interior minister to the executive directory (1 July 1798, 13 messidoro VI, n. 3872), ASMi, Popolazione p.a., b. 16.

22 'Legge relativa alla nuova divisione de' Circondari esterni del Comune di Milano' (22 December 1797, 2 nevoso VI), in *Leggi*, 1/I, 46.
23 Report from the Interior minister to the executive directory (1 July 1798, 13 messidoro VI, n. 3872), ASMi, Popolazione p.a., b. 16.
24 Session of the Great Council (2 June 1798, 14 pratile VI), *Redattore*, VI/155, 449–458.
25 Session of the Great Council (18 November 1797, 28 frimaio VI), *Redattore*, VI/27, 431–432; *Redattore*, VI/28, 435–437.
26 Session of the Great Council (7 April 1798, 18 germinale VI), *Redattore*, VI/35, 558–559.
27 Report from the Interior minister to the Executive Directory (17 October 1797, 26 vendemmiale VI), ASMi, Popolazione p.a., b. 16.
28 Session of the Great Council (7 February 1798, 19 piovoso VI), *Redattore*, 33/VI, 516–521. On Latuada: Vittorio Criscuolo, "Latuada Felice," in *Dizionario Biografico degli Italiani*, vol. 64 (2005): 44–46.
29 Unfortunately, the text of the second part of Dandolo's bill is not contained in the minutes of the Council of Juniors. Session of the Great Council (23 April 1798, 4 fiorile VI), *Redattore*, 66/VI, 1048–1051. On Dandolo: Paolo Preto, "Dandolo, Vincenzo," in *Dizionario Biografico degli Italiani*, vol. 32 (1986): 511–516.
30 Session of the Great Council (27 May 1798, 8 pratile VI), *Redattore*, 144/VI, 279–284.
31 Session of the Great Council (28 May 1798, 9 pratile VI), *Redattore*, VI/146, 312–316.
32 Session of the Great Council (30 June 1798, 12 messidoro VI), *Redattore*, VI/218, 1467–1470.
33 Session of the Great Council (23 August 1798, 6 fruttidoro VI), *Redattore*, VI/347, 1389–1390.
34 Session of the Great Council (28 August 1798, 11 fruttidoro VI), *Assemblee*, VII, 370–360.
35 Session of the Great Council (29 August 1798, 12 fruttidoro VI), *ibid.*, 401–404.
36 'Determina del Direttorio Esecutivo' (2 February 1799, 14 piovoso VII), in *Raccolta*, 1/VII, 224–225.
37 Announcement by the municipality of Milan (26 August 1798, 9 fruttidoro IV), *Raccolta*, 1/I.
38 'Legge sulla Guardia Nazionale' (11 August 1798, 24 termidoro IV), *Raccolta*, 1/V, 234–246.
39 Minutes of the debate on conscription, *Raccolta* 1/VI, 157–158.
40 Law (21 December 1798, 1 nevoso VII), *Raccolta*, 1/VII, 175–176.
41 Announcement by the municipality of Milan (21 January 1797, 2 piovoso V), in *Raccolta*, 1/II, 222–223.
42 Announcement by the General Police minister (14 January 1798, 25 nevoso VI), in *Raccolta*, 1/IV, 120–122.
43 Announcement by the General Police minister (6 August 1798, 29 termidoro VI), in *Raccolta*, 1/V, 250–251.
44 'Avviso dell'Amministrazione Dipartimentale d'Olona' (28 June 1800, 9 messidoro VIII), in *Raccolta*, 2/I, 25–26.
45 'Avviso del Comitato di Polizia d'Olona' (30 June 1800, 11 messidoro VIII), in *ibid.*, 26–27.
46 'Regolamento per la Polizia generale' (27 March 1802), in *Bollettino*, 1802, 41–47.
47 See, for instance, Vincent Denis and Pierre-Yves Lacour, "La logistique des savoirs. Surabondance d'informations et technologies de papier au XVIIIe siècle," *Genèses* 102 (2016): 107–122.
48 Denis, *Une histoire de l'identité*, 23–24, 30–31, 115–126.
49 See Noiriel, "L'identification des citoyens," 3–7.

50 'Legge relativa ai Registri' (24 July 1797, 6 termidoro V), *Raccolta*, 1/III, 88–91.
51 Report from the minister of Worship to the vice-president (11 July 1802), ASMi, Popolazione p.m., b. 51. On this regard, Bovara quoted a booklet published by the former member of the French Directory Louis-Marie de La Révellière-Lépeaux, *Réflexions sur le culte, sur les cérémonies civiles et sur les fêtes nationales* (Paris: Chez H.J. Jansen, l'an 5me [1796]).
52 Report from the minister of Worship to the vice-president (28 January 1803), ASMi, Popolazione p.m., b. 51.
53 'Osservazioni al regolamento' (undated, but after 09/1805), ASMi, Popolazione p.m., b. 2.
54 Draft decree (undated), ASMi, Popolazione p.m., b. 8. The declared aim of the minister was to equate Italian legislation with its imperial counterpart: draft letter from the minister of Justice to the viceroy (20 November 1812), *ibid.*
55 Draft letter from the minister of the Interior to the foreign minister (16 February 1807, n. II/787); letter from the justice minister to the minister of the Interior (19 March 1807, n. II/2271) both in ASMi, Popolazione p.m., b. 3.
56 Letter from the State Council's committee on municipal administration to the viceroy (13 May 1807, n. 294), ASMi, Popolazione p.m., b. 5.
57 Draft letter from the minister of the Interior to the foreign minister (7 August 1812, n. II/22321), ASMi, Popolazione p.m., b. 8.
58 Olivier Faron and Cyril Grange, "Une recensement parisien sous la Révolution. L'example des cartes de surete de 1793," *Mélanges de l'Ecole francaise de Rome. Italie et Mediterranée* 2 (1999): 795–826; Vincent Denis, *Une histoire de l'identité*, 32.
59 'Provvidenze per la polizia della Città di Milano' (20 November 1796, 30 brumaio V), in *Raccolta*, 1/II, 147–148.
60 Announcement by the General Police minister (14 January 1798, 25 nevoso VI), in *Raccolta*, 1/IV, 120–122.
61 Vincent Denis, *Une histoire de l'identité*, 153–154.
62 Session of the Great Council (18 November 1797, 28 frimaio VI), *Redattore*, VI/27, 431–432; *Redattore*, VI/28, 435–437. On La Hoz: Emanuele Pigni, "La Hoz, Giuseppe," in *Dizionario Biografico degli Italiani*, vol. 63 (2004): 86–88.
63 Draft law on passports (26 June 1798, 8 messidoro VI), *Assemblee*, V, 746–748.
64 Session of the Great Council (26 June 1798, 8 messidoro VI), *Assemblee*, V, 762.
65 Session of the Great Council (6 July 1798, 18 messidoro VI), *Redattore*, VI/237, 1762.
66 Announcement by the General Police minister (6 August 1798, 29 termidoro VI), in *Raccolta*, 1/V, 250–251. In the previous months, Brunetti had taken a decidedly pro-French stance that differentiated him from other Cisalpine élite actors: Rita Cambria, "Brunetti, Vincenzo", in *Dizionario Biografico degli Italiani*, vol. 14 (1972): 585–588.
67 'Regolamento per la Polizia generale' (27 March 1802), *Bollettino*, 1802, 41–47. On the *cartes de sûreté* in the first decade of the nineteenth century, see Vincent Denis, *Une histoire de l'identité*, 33.
68 Circular from the national director of the Police (31 May 1809), ASBo, Prefettura, 1809, tit. XX, rub. 10.
69 Session of the Legislative Council (14 February 1811), ASMi, Registri, vol. 642, 57r–74r. On the general population list, see Andrea Schiaffino, "Il ruolo generale della popolazione nell'esperienza storica del Regno Italico", in *Le fonti della demografia storica in Italia Volume 1. Parte 1*, ed. Andrea Schiaffino (Roma: Comitato italiano per lo studio dei problemi della popolazione, 1971), 519–541.
70 'Decreto sull'organizzazione definitiva de' cancelliere del censo' (29 June 1809), *Bollettino*, 1809, 203–214.
71 Session of the Legislative Council (21 February 1811), ASMi, Registri, vol. 642, 83r–85v.

72 'Decreto con cui viene regolato l'esercizio della polizia amministrativa sui forestieri che viaggiano nel regno, e sui nazionali che viaggiano all'estero, o girano nell'interno' (11 June 1811), in *Bollettino*, 1811a, 569–585.
73 Circular from the general director of the Police (18 January 1812, n. 501), ASBo, Prefettura, 1812, tit. XX, rub. 10 (parte 1). No provisions were mentioned for Italian internal passports in the French Empire, probably creating an asymmetry between the rights of movement of Italian and French citizens.
74 Draft law on passports (26 June 1798, 8 messidoro VI), *Assemblee*, V, 746–748.
75 Circular from the ministers of justice and the police to the central *dicasteri* and police sections (21 April 1801, 1 fiorile IX), ASMi, Miscellanea storica, b. 53.
76 Circular from the general director of the police (18 January 1812, n. 501), ASBo, Prefettura, 1812, tit. XX, rub. 10 (parte 1).
77 Circular from the interior minister to the bishops (14 September 1801, 27 fruttidoro IX, n. 47278), ASMi, Popolazione p.m., b. 1.
78 Letter from the interior minister to the finance minister (24 February 1802, n. 17927); letter from the finance minister to the interior minister (24 February 1802, n. 535); circular from the interior minister to the bishops (27 October 1802), all in *ibid.*
79 Conscription law (30 October 1801, 8 brumaio X), in *Raccolta*, 2/III, 131–132.
80 'Legge generale di Finanza per gli anni 1805 e 1806' (17 July 1805), in *Bollettino*, 1805, 425–447.
81 Ivana Pederzani, *Un ministero per il culto. Giovanni Bovara e la riforma della Chiesa in età napoleonica* (Milano: Franco Angeli, 2002), in particular 13–61 and 320–332.
82 This proposal is contained in at least two different yet similar undated bills signed by Bovara: 'Piano di legge sopra il contratto matrimoniale' and 'Progetto di legge sopra il contratto matrimoniale' both in ASMi, Popolazione p.m., b. 51.
83 'Titolo quarto del progetto del Codice Civile per la Repubblica Italiana' (undated), *ibid.* At the end of 1803, the author of the draft Alberto De Simoni submitted a new draft Code in which civil marriage was completely managed by the civil authorities: Piergiorgio Peruzzi, *Progetto e vicende di un Codice Civile della Repubblica Italiana (1802–1805)* (Milano: Giuffrè, 1971), 8–193, 246, 309.
84 Letter from the minister of Worship to the vice-president (28 January 1803), ASMi, Popolazione p.m., b. 51.
85 Report from the minister of Worship to the vice-president (28 January 1803, *ibid.*
86 Report from the Justice Committee to the Legislative Council (undated), ASMi, Registri, vol. 597, 168–181.
87 Session of the Legislative Council (8 March 1803), *ibid.*
88 In the new formulation of the law, priests could be delegated as "assistants" to manage the civil registers. The choice between a municipal official and a priest remained possible for couples wishing to record their marriage in the civil register, and the formal distinction between the civil registration of marriage and the religious rite was confirmed. The only significant change affected the other religious minorities, who were not allowed to have a specific religious delegate: confidential letter from the minister of Worship to the secretary of state (24 February 1804, n. 1945/I), *ibid.*; 'Progetto di Legge sopra il Contratto Matrimoniale' (24 February 1804), ASMi, Popolazione p.m., b. 51.
89 Session of the Legislative Council (7 April 1804), ASMi, Registri, vol. 602, 22–27. The discussion over Bovara's bill opened on 3 April and went on until 16 April.
90 Copy of letter from the minister of Worship to the justice minister (6 July 1804, n. 6929), ASMi, Popolazione p.m., b. 51.
91 The correspondence between the three ministers is in *ibid.* The vice-president Melzi d'Eril agreed with the proposal in principle. Nevertheless, he could not issue a decree on such a matter while the Concordat was still being implemented:

letter from the minister of Worship to the interior minister (12 September 1804, n. 9936/I), *ibid.*

92 Session of the State Council (13 May 1805), ASMi, Popolazione p.m., b. 5.

93 *Codice Civile di Napoleone il grande pel Regno d'Italia. Quarta Edizione originale ed officiale. Tomo I* (Milano: Dalla Reale Stamperia, 1806), 12–33.

94 Confidential letter from the minister of Worship to the minister of internal affairs (26 August 1805, n. I/9628), ASMi, Popolazione p.m., b. 1.

95 Session of the Legislative Council (12 March 1806), ASMi, Registri, vol. 611, 52r–58r.

96 Session of the Legislative Council (13 March 1806), *ibid.*, 59r–65r.

97 Session of the Council of State (18 March 1806), ASMi, Registri, vol. 609, 83r–90r.

98 'Regolamento per l'attivazione in tutto lo Stato dei Registri delle nascite, de' matrimonj e delle morti' (27 March 1803), *Bollettino*, 1806a, 168–223.

99 Letter from the interior minister to the justice minister (6 October 1806, n. II/11801), ASMi, Popolazione p.m., b. 4.

100 Confidential letter from the minister of Worship to the interior minister (13 October 1806, n. I/11258), *ibid.*

101 Letter from the interior minister to the minister of Worship (22 October 1806, n. 12637), *ibid.*; circular from the minister of Worship to the archbishops and bishops (28 October 1806, n. I/11743), ASMi, Popolazione p.m., b. 2 and 4.

102 Report from the interior minister to the viceroy (21 December 1807, n. 19569), ASMi, Popolazione p.m., b. 5 and 6; draft decree of the interior minister (21 December 1807), ASMi, Popolazione p.m., b. 4 (draft) and 6. Di Breme's bill was based on a proposal and a report drafted by his subordinate Benedetto Bono, general director for the administration of municipalities: draft report from the general director of the Municipal Administration to the interior minister (18 December 1807, n. 1734); draft decree of the general director of the Municipal Administration (18 December 1807), both in ASMi, Popolazione p.m., b. 5.

103 Session of the Legislative Council (8 January 1807 [*recte* 1808]), ASMi, Registri, vol. 627, 5r–6r; letter from the president of the Legislative Council to the secretary of state (13 January 1808, n. I/7), ASMi, Popolazione p.m., b. 5.

104 Report from the justice minister to the viceroy (16 February 1808, n. III/1185), *ibid.* Some weeks earlier, Luosi had received an internal report expressing similar views: internal report to the justice minister (30 January 1808, n. III/1185), *ibid.*

105 Confidential letter from the minister of Worship to the secretary of state (26 March 1808, n. 2191), *ibid.*

106 Session of the Legislative Council (28 April 1808), ASMi, Registri, vol. 627, 162r–166v; session of the Legislative Council (2 May 1808), *ibid.*, 171r–173r.

2 Policing Identities

Identity Documents in an Urban Context

When the Italian identification system was extended to the Venetian territories in 1806, Vicenza was a small city of approximately 25,000 inhabitants on the busy road that connected Venice to Milan (Figure IV).[1] The city was still surrounded by walls, which played a crucial role in defending the urban population during the 1809 uprisings. The urban structure was made up of different layers marked by the famous Palladian Renaissance palaces. In recent years, however, various local administrations had sought to rationalise the city's layout. They demolished gates in the walls, straightened out roads, and installed a first system of public lighting. Moreover, both the Habsburg and Napoleonic administrations had started implementing a gradual closure of several churches and monasteries.

Vicenza was an industrial city. At its peak, the silk industry occupied 3,550 workers, of which almost one-third were women and one-fifth were children. By 1806, though, this remarkable manufacturing activity was traversing a moment of profound crisis. During the ten years of pro-French government, the economic policies of the Napoleonic system led to a dramatic drop in the number of people employed in the silk industry. Unemployment and vagrancy were on the rise, while the religious welfare institutions found it ever more difficult to provide assistance to the poor.

After belonging to the Republic of Venice for almost four centuries, the city had experienced a brief pro-French period in 1797, when a new "democratic" government had replaced the traditional aristocratic and Venetian authorities. Vicenza's eventual annexation to the Habsburg Empire in 1797 led to the restoration of aristocratic government but did not interrupt attempts to rationalise urban society. The city's subsequent attachment to the Kingdom of Italy resulted in new institutional changes, marked by Vicenza's establishment as the capital of the Bacchiglione department. The local prefect, a state official external to the local community, appointed the new municipal administration.[2] At the same time, state-appointed judges began to administer the local justice system. Moreover, a central police office replaced the several forces that had carried out law enforcement in the previous centuries. The new police commissioner reported directly to the prefect. To control public order in the city and its immediate surroundings, he could rely on four

DOI: 10.4324/9781003412434-3

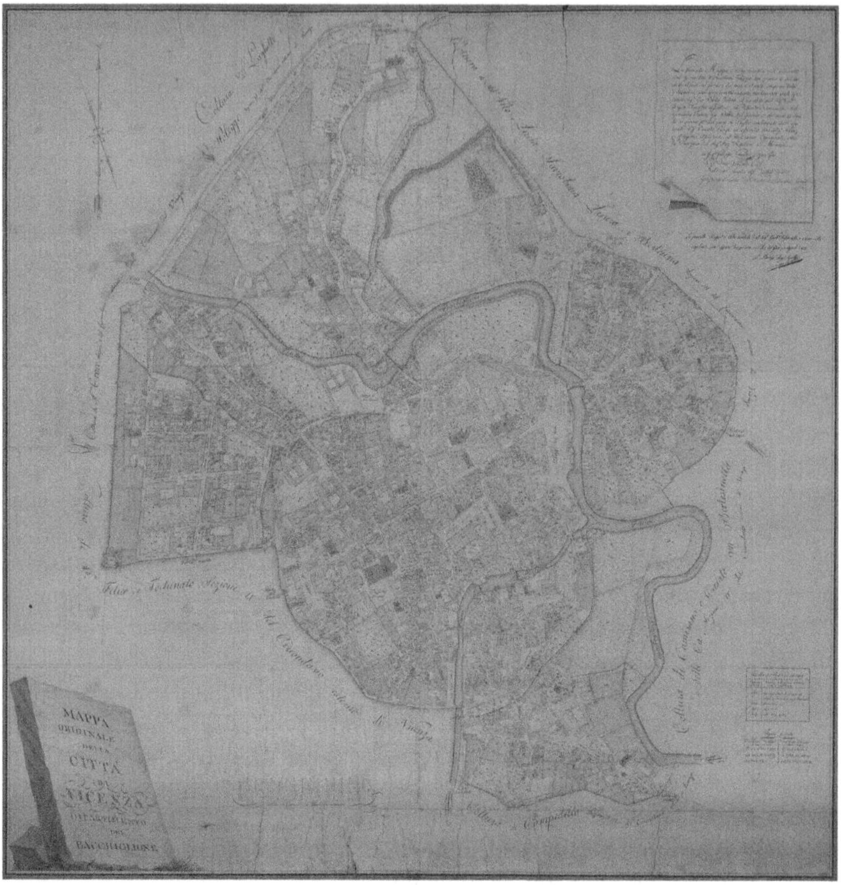

Figure IV Cadastral map of Vicenza in 1813 (ASVe, diritti, capo XXIX, cap. 2584, art. 3).

employees and a few guards.[3] This small police force was at the centre of the identification procedures introduced by the Napoleonic government. This chapter relies on these police archives to reconstruct how the local authorities implemented the idea of a legal identity managed by the state.

The local prefect introduced security cards to the territory of the department of the Bacchiglione – including Vicenza itself – in October 1806.[4] The practical purpose of these identity documents was to 'provide a simple tool for citizens to travel throughout the department and the kingdom with the appropriate documents'.[5] This wording was not dissimilar to that of the 1802 police law that had established security cards as mandatory documents for all travellers.[6] During the summer of 1807, however, new departmental provisions sought to radically change the use of security cards in two different ways. On the one hand, security cards became mandatory for all citizens – and not only for those who were travelling. On the other hand, the provisions specified that 'all citizens liable to be recruited' in the army had to obtain a security card.[7] Thus, the new rules explicitly excluded both young people and women from the identification system. The decree stated quite clearly that the new measures were connected to the conscription system. It seems more than likely that the political situation had provoked these changes. In the same months, in fact, the Venetian population was putting up a fierce resistance to the introduction of conscription.[8]

Moreover, the 1807 provisions specified that security cards were not to be issued to those who could not provide proof of good behaviour, means of subsistence, and a clear criminal record.[9] The departmental rules echoed the Cisalpine idea that idlers and suspects were individuals at the fringe of citizenship – and thus not reliable enough to be given an ordinary identity document. The strong connection between surveillance and identity documents was confirmed by the departmental announcement that explained the 1807 regulations to the general public:

> The police – the prefect stated – is an institution through which public order is maintained, crimes are avoided, criminals are apprehended, and life and the property of citizens are guaranteed. These healthy aims are mainly obtained by giving each honest individual a tool by which they may guarantee themselves in front of the authorities. For this purpose, security cards were established in the last year.[10]

In the words of the police commissioner, the security cards' aim became to 'guarantee honest citizens in their journeys and in their sojourns in this area'.[11] Local authorities saw the new identification system as a tool with which to tackle two of the main problems that they were facing: delinquency and resistance to conscription. Security cards – and through them, legal identities – were strictly intertwined with a desire for social control. But first, they had to take on a material form.

Material Identities

The first step towards the creation of legal identity documents was to print them. The authorities pre-printed registers in which every page contained two copies of the same security card. One remained with the authorities, who gave the other one to the applicant. The issuing procedure, thus, did not just generate an identity document, but also a register of all previously issued cards.[12] Over the years, the overall appearance of security cards remained almost the same: they were rectangles of heavy white paper on which a form was printed in black ink (Figure V). What really distinguished one security card from another was the information added to the pre-printed form. Every type of security card contained an indication of what its holder was allowed or not allowed to do. A standard security card, for instance, read

> [the holder], identified by the undersigned witnesses, is guaranteed by the present [card], in front of the civil and military authorities, for travel throughout the territory of the Kingdom of Italy.

Every security card mentioned its place and date of issue, together with basic information on the holder (name, surname, place of birth, place of residency, profession, age). Furthermore, the cards contained a physical description of the holder, providing information on the colour of their hair, eyebrows, eyes, and beard (if they had one), along with a description of the shape of their nose, mouth, and chin. Height was also specified, together with the presence of any 'visible marks'.

However, the security cards were not uniform in their appearance. Every single department had a slightly different layout that could also change over time. In the department centred around Vicenza, the layout changed at least three times and the physical description of the holder was moved to a separate section in 1807. This feature – that was maintained in subsequent years – cannot be found in the security cards issued by other departments.[13] For example, a security card issued by the department of the Tagliamento in January 1807 contained several elements that cannot be found in the cards issued at the same time by the department of the Bacchiglione: a coat of arms of the Kingdom of Italy was printed at the top of the card; the issuing municipality was specified just under the name of the department; and the prefect's stamp was substituted by a written formula.[14] Every security card had a serial number in one of the top corners. Here again, a lack of uniformity can be found between cards issued in different departments and in different periods. For instance, in the department of the Bacchiglione, this number was always handwritten, except during a short period in 1807. By contrast, some security cards issued by the department of the Tagliamento and by the Venice police commissioner had their serial numbers impressed.[15]

This lack of uniformity implied a certain degree of flexibility on behalf of the local bureaucrats and police. In April 1808, the 18-year-old Bortolo Rocco was arrested in Vicenza while in possession of an irregular document. The young man had a security card issued by the municipality of Feltre, in the department of the Piave. The problem was that his security card was 'handwritten on stamped paper which included the legal seal'. For unknown reasons, his municipality had provided him with a handwritten security card. After his arrest, however, he was released with a simple endorsement on his (handwritten) security card.[16] The tolerance level for irregular documents issued by local authorities was understandably high.

These differences in the layout show how one of the central authorities' primary aims – namely, the promotion of uniformity in identity documents throughout the state – had not been achieved, even several years after the law of 1802 had originally been implemented. The reasons underlying this lack of uniformity remain unclear. While it is true that the printing of these documents was delegated to the departments, it is also true that the central authorities could have easily imposed a uniform layout on their peripheral subordinates. If the reasons for this lack of uniformity remain unclear, its consequences are easy to imagine: facing several different layouts, police forces could hardly have been able to detect forged security cards. The

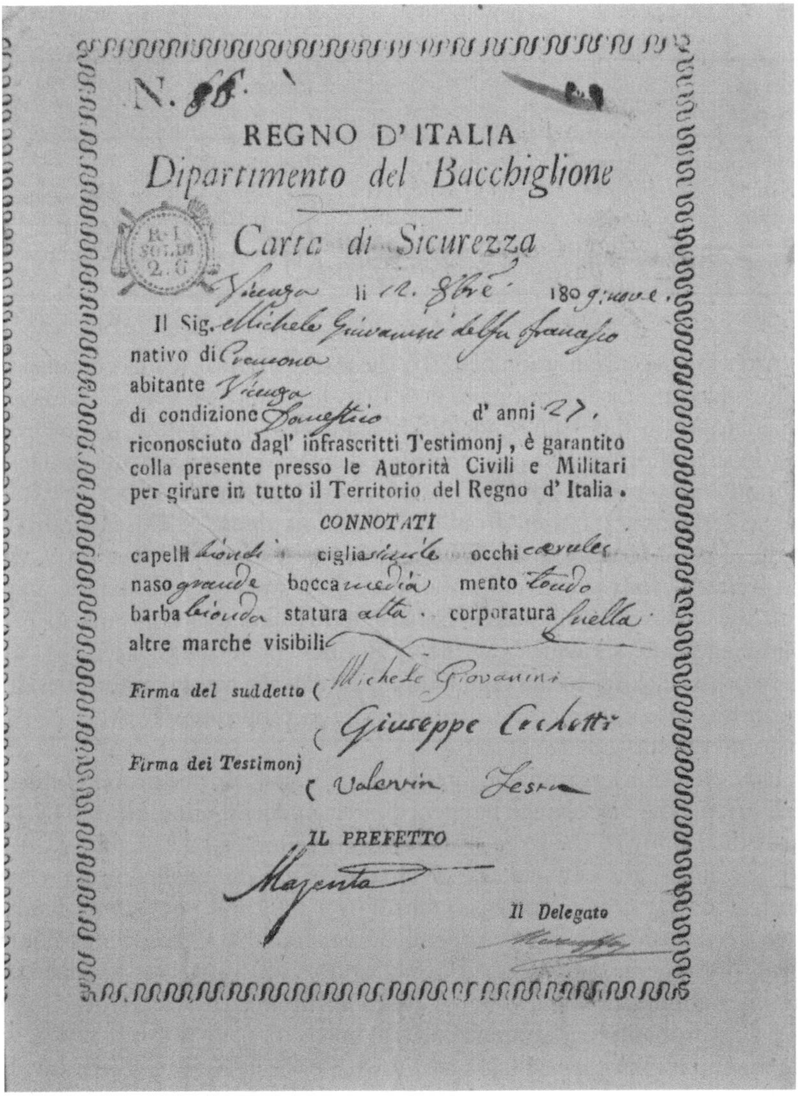

N. 66

REGNO D'ITALIA

Dipartimento del Bacchiglione

Carta di Sicurezza

Vicenza li 12. 9bre 1809

Il Sig. Michele Giovannini del fu Francesco

nativo di Cremona

abitante Vicenza

di condizione Domestico d' anni 27.

riconosciuto dagl' infrascritti Testimonj, è garantito colla presente presso le Autorità Civili e Militari per girare in tutto il Territorio del Regno d' Italia.

CONNOTATI

capelli biondi ciglia oscure occhi cerulei

naso grande bocca media mento tondo

barba bionda statura alta corporatura snella

altre marche visibili

Firma del suddetto (Michele Giovannini

Firma dei Testimonj (Giuseppe Cecchetti (Valentin Zesca

IL PREFETTO

Mazenta

Il Delegato

Figure V Ordinary security card issued in 1809 (ASVi, Commissariato, b. 13, f. 65).

security cards' lack of uniformity was precisely one of the reasons that led the authorities to introduce uniform registration certificates throughout the state in 1811 – almost ten years after the first introduction of a coherent documentation system.

This indeterminacy was also reflected in decisions regarding which authorities were supposed to issue the security cards. In less than three years, the issuing authorities in the department of the Bacchiglione changed several times (Table I).

Table 1 Changes in the issuing authorities of security cards in the department of the Bacchiglione

Area of jurisdiction	*1806*[18]	*1807*[19]	*1808*[20]	*1809*[21]
Department	Prefecture	Prefecture	Prefecture	
District		Vice-prefectures		
Canton	Cantonal authorities	Justices of the peace / police commissioners	Vice-prefectures	Municipalities / police commissioners

After the decentralisation of 1807, the task was once again centralised the following year. It is interesting to note that this choice was aimed at avoiding irregularities in the issue of identity documents. The prefect acknowledged that local authorities had previously not displayed the needed 'discipline and caution' in the issuing procedures. This levity had made it impossible to distinguish 'the honest citizen' from 'the men of ill repute'.[17] This centralisation of the issue of security cards, however, was reversed in October 1809, when the prefect – 'for the convenience of the inhabitants' – entrusted the mayors with the task.[22] The new decentralised status quo was then upheld in the following years, showing how the authorities' preferred to keep the cards' point of issue closer to the population. Just like the central authorities, local authorities also apparently struggled to find a compromise between convenience and the surveillance of citizens.

However, other reasons might also be assumed for this process of decentralisation. The last change happened after a general annulment of security cards that followed the large-scale revolts of summer 1809. Contrary to the explicit aim – 'the convenience of inhabitants' – the change was probably motivated by a need to increase control over the rural population. It would have been easier for the peripheral bureaucracy to exercise this control, since, presumedly, it possessed a higher level of awareness regarding the 'moral and political behaviour' of rural inhabitants. However, even though this was probably true, municipal administrators often sympathised with the inhabitants of their villages and towns, even if these had been involved in the riots.

This, for instance, was the case in the small village of Gazzo. Here, the mayor issued security cards to three men that had been involved in the 1809 revolt. Moreover, the choice to decentralise the issue of security cards led to inefficiencies, especially in smaller municipalities. An inspection of the card registers in the village of Gazzo itself revealed that the municipal bureaucrats had made several mistakes and committed a number of irregularities in the issuing procedures.[23] Still, the departmental authorities never applied any form of systematic control over the ways in which municipalities issued the security cards, and merely intervened in individual cases in which problems emerged. In fact, systematic control would probably have been too expensive for the weak state authorities. Precisely because municipal officials were more attuned to their local communities, they occasionally acted as biased mediators in identification procedures.

Once printed and delivered, security cards were ready to be filled out by the issuing authorities. However, applicants had to fulfil some conditions. First of all, they had to prove some sort of relation with the place in which they were requesting the identity documents. Neither the national nor the departmental provisions mentioned a residency requirement for the obtention of a security card. But at the departmental level, a ruling specified that mayors could provide cards to the 'inhabitants' of their municipalities.[24] This criterion, however, was never decisive for the local police. In December 1809, the commissioner forced a hatter to return to his birthplace to get a security card, even if he was living in the departmental capital.[25] In May 1810, the same police commissioner denied a security card to a citizen who was legally domiciled in the city 'because he is not from this municipality'.[26] The prefect of the Bacchiglione then felt obliged to reiterate to the police commissioner that 'the ordinary security card can be given to all the people born or also simply domiciliated in the municipality'.[27]

However, having a strict and lasting relation with the place of issue was not in itself sufficient to obtain an identity document. Being born in Vicenza was not enough for Giuseppe Scorzato, who returned to the city in December 1807 after two years of absence in order to get a security card there. After a first request, he had to wait for a certificate of domicile to come directly from Mantua, where he had lived for the previous two years.[28] Being legally domiciled in the city could also not be sufficient if the applicant had only moved there recently. In November 1808, a young librarian was arrested for not having a security card in Vicenza, where he had lived and worked for one year. In his interrogation, he explained that he had tried to get a security card in Vicenza several times, but that the authorities had always refused his request 'since I am not from Vicenza'.[29]

The residency requirements for obtaining an identity document were largely irrelevant to agents of the state. The approval – or the rejection – of an application for documents was not inspired by legal criteria such as the definition of domicile contained in the Civil Code.[30] On the contrary, what seemed crucial for an application's success was whether the applicant was tied to social networks in the city. In approving or rejecting an application for documents, bureaucrats assessed whether the applicant could or could not entertain effective and substantial relations with local society.

This last observation is also proved by the other main condition for the issue of a security card, namely, the presence of two witnesses. According to the first announcement made by the prefect of the Bacchiglione in 1806, the two witnesses had to be 'personally known' to the issuing authority. Moreover, they had to guarantee the applicant's 'probity, morality and good behaviour'. In September 1808, a departmental announcement repeated that the witnesses had to be 'honest and trustworthy [people]' and that they had to know the applicant personally.[31] In this way, the bureaucracy tried to involve social relationships in the identification procedure.

In practice, however, there were several cases in which witnesses and applicants did not know each other. During an interrogation, for instance, a prisoner could not remember the second witness who had signed his security card only six days previously.[32] Another hint of this practice comes from the records of the prefectorial police office of the department of the Reno. Here, the repression of brigandage in the countryside had provoked an unexpected surge in the requests for security cards in 1810.[33] One of the police commissioners charged with issuing the cards therefore resorted to questionable short-cuts in the procedure, especially as far as regarded witnesses. He wrote to the prefect of the Reno that the applicants were mostly illiterate and could not act as witnesses for each other. Consequently, the commissioner had to ask 'educated people that were casually in the office' to serve as witnesses – and they had done so for several hours.[34] The resort to witnesses who did not know the applicants was probably a common practice, even in ordinary situations.

As time moved on, witnesses were no longer required to vouch for an applicant's good behaviour.[35] However, applicants still had to prove their good reputation, even through the use of written evidence. As of 1807, they had to present a certificate of good behaviour written by their parish priest and countersigned by their local municipality.[36] In 1808, a clean criminal record issued by the local authorities became another requirement for the obtention of a security card.[37] All these precautions were meant to make the process by which security cards were issued more reliable. In fact, the issue of state identity documents could not avoid being based on oral and written certifications of the applicants' social reputation. The security cards provided not only a material embodiment of an individual's identity, but also their (certified) reliability.

In practice, these precautions did not completely prevent mistakes and misattributions. In September 1809, for instance, the police in Brescia arrested a 23-year-old deserter with a valid security card issued by the Vicenza police office. The prefect rebuked the police commissioner for having issued a security card to a young foreigner: his age and provenance were warning signs that should have inspired more caution.[38] The police commissioner's defence was meaningful: he had not received any complaints about this young man, and his local criminal record was clear.[39] In this case, good reputation was enough for a recruitable foreigner to obtain a security card.

If all the conditions were met and the applicant was in the right office, the issuing authority would then fill out the security card. An official would put an official stamp on the identity document and ask the applicant to pay the required fee.[40] He would write down all the personal information that he could glean from parish certificates, such as age and place of origin. Special attention was paid to the physical appearance of applicants. The security cards stored in the Vicenza police commissioner's archives suggest that there were no precise guidelines on how to write this physical description. The bureaucrats enjoyed great freedom when filling out these fields. Nevertheless, they inevitably ended up adhering to a limited set of adjectives for each field (Table II).

The single fields, taken separately, appear to be quite general. The definition of a 'regular' nose, for instance, appears to be extremely arbitrary and unstable. Combined together, however, these single fields allowed for cross-checks that could easily reveal any instances of misappropriation or forgery. Moreover, they associated a physical description to a name and surname, creating a more detailed and certified identity for the holder.

Finally, to be valid, the identity cards had to be signed by five different people: the holder, the two witnesses, and two representatives of the issuing authorities.[41] This full-blown storm of signatures spurred the authorities themselves to look for simplification strategies. The prefect of the Bacchiglione, for instance, started to "sign" with a stamp of his signature at least from 1809. The applicants and the witnesses could, on the other hand, be illiterate. In most of these cases, officials wrote down 'illiterate' in lieu of the requested signatures. However, they could also find more creative solutions. In 1808, for instance, the prefect of the Bacchiglione had to annul several cards issued by a police employee in Vicenza, since the latter had written the witnesses' signatures with his own hand, 'without even calling [the witnesses] back [to the office] to sign'.[42] All the precautions established by the authorities were useless if the administrative personnel found them hard to implement or easy to avoid. It might be presumed that, as more precautions were added, the issuing procedures became more and more time-consuming, leading local bureaucrats to cut corners to avoid what they perceived to be a waste of time.

Once all the identity cards in a register had been issued, officials compiled alphabetical lists of all the applicants from each register 'in order to easily find [them] in case of need'.[43] This tool allowed the local authorities to potentially have access to a remarkable database containing basic information on applicants and their appearance. Every month, each issuing authority in the department had to report how many cards they had issued along with their serial numbers.[44]

This procedure theoretically created a set of legal identities combined with physical descriptions and attached to a material support, the security card. Thanks to the stubs and the alphabetical lists, the authorities could mobilise this huge amount of information every time they needed it. However, as we have seen, the unreliability of the officials who actually issued the cards undermined the potential reliability of these legally created identities.

Table II Physical descriptions taken from ordinary security cards stored in ASVi, Commissariato

Hair	*Eyelash*	*Eyes*	*Nose*	*Mouth*	*Chin*	*Beard*	*Height*	*Physique*
Black	Black	Cerulean	Broad	Big	Large	Chestnut	Hugh	Slender
Chestnut	Chestnut	Black	Regular	Medium	Long	Blonde	Ordinary	Big
Brown	Brown	Chestnut	Big	Ordinary	Round	Short	Big	Substantial
Dark	Dark	Blue	Large	Small	Oval	Black	Small	Ordinary
Blended	Blended	Dark	Small		Short	Dark		
Reddish	Reddish	Brown				Grey		
Blonde	Blonde					Brown		

The Napoleonic procedures of identification were still dependent on mediators who were not completely external to the communities that they had to identify.

The Identity of the Suspect

Not every identity was equal before the law. Certain individuals – and certain identities – were singled out for special treatment by the authorities. The 1802 police law gave a precise definition of who was to be considered a suspect: idlers and vagrants, beggars who were fit for work, former prisoners, and people under police investigation. The law allowed legislators to issue further regulations on this matter, including punishments for transgressors of up to one year in prison.[45] The definition of who was a suspect was broadened, in practice, over the following years. For instance, when the Vicenza police commissioner communicated a list of tasks to his officers in 1807, he mentioned a larger definition of who could be considered a suspect:

> [The police officers'] zeal must be directed towards the behaviour of the idlers, vagabonds, or otherwise suspect people, without neglecting those who habitually spend time in the cafés, inns, and other similar public places, or [those who] lead a wasteful life, and who are known to be lacking the commensurate means of subsistence. [...] Special surveillance must not avoid children neglected by their relatives, well-known procurers and petty thieves, crooks, those who fence, or those who are suspected of being second-hand traders and stockpilers of stolen goods and similar [things].[46]

Over the years, the definition of suspects had expanded until its boundaries became quite blurred. Thanks to this vagueness, police forces enjoyed a large degree of arbitrariness in the definition of who was a suspect, which in turn led to suspicion becoming a crucial parameter in police practices.

Suspects had to be listed at every level of the administration, in a way that was not so dissimilar to the contemporary compilation of lists of notables. These registers of suspects were compiled, at first, at the municipal level. They contained basic information on local suspects and on the reasons for which they had been assigned this status.[47] The vice-prefectures later started to collect all the municipal registers in order to create lists of the suspects present in their districts, which were then sent every month to the prefecture.[48] Finally, since the police office of the prefecture could issue security cards for the whole department, a departmental register of suspects was established in September 1808.[49]

In Vicenza, a first list of suspects was drawn up in December 1807, when the prefect of the Bacchiglione ordered the newly appointed police commissioner to arrest all of the suspects who were present in the city.[50] The commissioner passed on the task to the chief of the police guards, who, one day later, delivered a handwritten 'Note of the idlers, thieves, suspects and criminals, lacking means of subsistence'.[51] The list contained the names, surnames, nicknames – and sometimes the places of birth – of 68 men who were domiciled in the city. 39 of them were listed as 'thieves', while 22 were 'suspects'.[52] The list reveals some interesting hints of how a common police agent conceived the idea of suspicion. The frail distinction between 'thieves' and 'suspects' shows that, at the local level, suspicion had still not been formalised as a category with precise parameters – even if the title of the list itself recalled the four categories mentioned in the 1802 police law.

Physical appearance played an important role in the distinction between reliable and suspect individuals. Clothing was particularly important for the police. Some traces of this attitude can be found in some of the rules issued by the authorities. According to regulations regarding the surveillance of the city gates, citizens and foreigners passing through the gates had to undergo a strict inspection and registration process. If caught without a valid document, they were to be brought to the police commissioner for further investigation. However, the regulation prescribed an exception for 'distinguished people and soldiers, especially if high-ranking': in this case, declaring their name and surname was enough to pass through the gate.[53] Apparently, wearing an officer's uniform or high-quality clothing was a sufficient guarantee for the state, even in the absence of a valid identity document. The same criteria did not apply to religious clothing, as this was probably easier to obtain. In February 1810 – a few months after the imprisonment of pope Pius VII – a national provision ordered that anyone wearing religious clothing who could not certify their actual identity was to be considered suspect.[54] Thus, in line with the old Italian saying, Napoleonic bureaucrats also understood that *'l'abito non fa il monaco'*: a habit does not make a monk.

However, clothing was not the only element of a person's physical appearance that could provoke suspicion. In April 1812, for instance, the gendarmerie arrested the 34-year-old peasant Bonifazio Cormellotto because he had no identity card. In their report of the arrest, the gendarmes wrote: 'Having considered [Comerlotto] as a suspect person, also because he had a military haircut, we felt compelled to arrest him.'[55] In his interrogation, Cormellotto explained his haircut to the puzzled police commissioner: 'Although you see me with a shaved head, you must not suspect that I am a deserter, because this [haircut] is a habit of mine.'[56] An unusual haircut was indeed sufficient to attract the attention of the police force, if it could be related to desertion. Here, as elsewhere, police procedures were inspired by a large degree of arbitrariness.

The category of suspicion was applied to places too. In particular, the authorities focused their attention on establishments in which foreigners found temporary accommodation. Over the years, the local authorities issued several sets of rules geared to obtain updated information from hoteliers and innkeepers about their guests.[57] In November 1807, the prefect ordered the police commissioner to accurately monitor certain commercial activities

> I must also order you to carry out, starting from tonight, accurate inspections in the hotels and inns that are individually more suspect, and to arrest any individual housed there who does not have legal documents to justify their identity.[58]

As we will see below, this recommendation led to the frequent night raids being carried out in inns, taverns, and other 'suspect places'. The first of these raids happened the following night, when the police guards arrested 'ten individuals, most of whom are suspect because they are already known to the justice system for being shoplifters'.[59] In June 1811, the report of an arrest started with these words: 'last night during the inspection of the inns and suspect places [...]'.[60] The concept of 'suspect places' had become a standard formula that did not require any further specification.

As we have seen, these regulations envisioned the security card as a way of certifying who was a good citizen in the eyes of the law. One of the implications of this idea was the need for a clear and material distinction between "good" and "bad" citizens. Special identity cards for suspect, the so-called *module IV* and *VII*, were the practical solution by which this aim was achieved. The Cisalpine police law of 1797 had already allowed for control over the movements of suspects and other types of 'lowlife'.[61] Moreover, the 1802 police law prescribed 'special surveillance' for suspects.[62] From the same year, the legislation required all suspects to notify their place of residence. Suspects could change their place of residence only if authorised by the authorities. If they did so, they received a 'special card' allowing them to move to their new home. The same decree created a document for those who were released from prison and wanted to be domiciled in a new municipality. The local police would then give the suspect a card that 'guarantees for their permanence in this jurisdiction'.[63]

A departmental regulation specified that being registered as a suspect would make it impossible to obtain a security card.[64] In September 1808, a departmental circular made the equivalence between suspicion and *module* even more automatic and unavoidable:

> From now on, all those people who are considered to be suspect, according to the definition contained in the second article of the law of 20 August 1802, shall be sent by the police commissioner to the prefecture with a report and the documents that are related to it, so that they may be provided with whichever *modula* pertains to them.[65]

From a formal point of view, the *modula VII* was a permission to stay in a certain municipality for people who had formerly been in jail or who 'live[d] a life of crime'. As we will see, however, this certificate was also used to strongly limit the holder's rights of movement. The *modula VII* included a request that the authorities keep the holder under 'strict surveillance', as they were 'suspect and prone to crime'.[66] On the other hand, the *modula IV* was used to allow suspect people to move from one place to another without a police escort.[67] The prefects and the vice-prefects were in charge of issuing both types of *modula*.[68] Even if it was the local police commissioner who proposed which penalties should be linked to each *modula*, their issue always remained under the jurisdiction of the political authorities. The surveillance of suspects was more of a political matter than a mere police task.

The *module*'s material incarnation was not too different from that of the ordinary security cards. They were almost the same size and followed the same pattern with a header at the top, information on the carrier in the middle and signatures at the bottom. Nevertheless, both types of *module* mentioned their discriminatory nature at the top of the page.[69] This was probably a useful expedient for immediately pointing out the "special" nature of the card to the authorities who were checking it. It is therefore not surprising that the *module* were rather bare in comparison with other types of security cards. No graphic element decorated either the header or their frame. Less information was contained, and the pre-printed form was definitely blunter. As was the case for the other security cards, the elements of the *module* could differ depending on which department had issued them – even if all the headers began with a printed 'Regno d'Italia'.[70] Moreover, for the *modula IV*, a different issuing authority provided the final signature in each department.

These documents did not need signatures from any witnesses to be considered valid. This detail is quite revealing of the substantial difference between *module* and ordinary security cards. While the latter were intended as a form of guarantee for the holder, the former were explicitly a form of control over people whom the authorities considered to be suspect. This discriminatory nature is explicit in the pre-printed text contained in the *module*. The pre-written form for the *modula VII* opened with 'it is accorded to', followed by the holder's information and by the concise formula 'the permission to stay in this city'. The text of the *modula IV*, on the other hand, only contained information on the holder plus the brief formula 'Is moving to the jurisdiction of ...'. The pre-written text ended the significant phrase 'The police of ... presents [the holder] to the police of ...'. Hence, the pre-printed text clearly marked the negative implication of holding a *modula* instead of a security card.

As has been mentioned above, most of the *modula IV* were issued for the purpose of transferring a suspect from one city to another without a police escort. This practice was particularly common for the deportation of suspects and formerly imprisoned people to their hometown or declared place of residence. For example, the young salesman Antonio Tessari was deported from the department of the Brenta to that of the Bacchiglione with a *modula IV* in January 1808. His arrival was announced by the prefect of the Brenta, who ordered that Tessari be kept under surveillance and provided all the physical information needed to identify him. When Tessari arrived in Vicenza, he was added to the list of local suspects and was probably given a *modula VII*.[71] Similarly, the itinerant cobbler Giacomo Favero was sent to his hometown of Belluno with a *modula IV* after he had been caught by the Vicenza police without an identity document.[72]

The functions of the *module* could, however, be surprisingly flexible. The generic nature of the pre-printed form, in fact, implied that it could be used in an almost infinite variety of ways, depending on the situation and the will of the issuing authorities. This was the case for the peddler Giovanni Crevini, who was given a peculiar *modula IV* by the authorities in Vicenza in January 1810. He had received a *modula VII* two years earlier, when he had been put under investigation for assault. Since the court had still not communicated the results of the investigation to the police office, this original *modula* was still valid. Moreover, his *modula* had been confirmed in December 1809, after he was arrested as a suspect in Brescia, probably because he did not have a security card. When he returned to Vicenza, Cravini asked for a security card for the purpose of travelling to Padua for business. At that point, the prefect found a compromise that satisfied both Cravini's need to travel and the authorities' will to keep him under surveillance. The solution was a *modula IV* that specified that it was only valid for one month and only for a trip, 'there and back', to Padua for business purposes.[73]

The *modula VII* was even more flexible, but it took some time for police forces to start fully exploiting its potential. When the *modula VII* was first introduced, the Vicenza police commissioner tended merely to add the name of suspects' final destination at the end of the pre-printed form, in order to preclude any further movement on their behalf. As time went by, however, the authorities annotated the *module VII* in increasingly complex ways. In March 1810, the baker Andrea Novello received a *modula* with this annotation:

> [The holder is given permission to stay in the city] with the obligation of renewing this [card] from month to month; of not wandering around one hour after sunset, in suspect places; and finally of committing himself to finding stable employment within twenty days. [The holder] is not to leave this municipality without superior permission, under the penalty of three months in prison in case of infringement.[74]

These annotations applied several hindrances to the baker's freedom of movement, and added the significant obligation to find a job in a very short period of time. In spite of their harshness, these kinds of provisions quickly became the norm. In July 1809, for instance, the prefect of Vicenza assigned a similar *modula VII* to a recently released inmate by simply writing 'give him an appropriate *modula VII* with the usual precepts'.[75] Albeit with some slight changes in content, this formula was repeated countless times in the following years.

The flexibility of the *modula VII* was also exploited in other ways. In late December 1811, complaints were filed against the hatmaker Francesco Monzerle

> for the constant arguments and fights that he provokes in Pietro Gnoato's inn in Pescaria, as a consequence of which several clients have declared that they will be forced to avoid going to that [inn] to avoid inconveniences if the aforementioned Monzerle continues to frequent it.[76]

In order to avoid further problems, the police commissioner decided to give Monzerle a *modula VII* with a precise injunction: '[Francesco Monzerle must] from now on avoid entering, staying or [doing anything] other in Pietro Gnoato's inn in Pescaria'.[77] Similarly, in June 1812, the doorman Antonio Boriero received a *modula VII* that stated

> the aforementioned Boriero is also forbidden from entering Anna Cavaliere's house under the penalty of three days [in prison] the first time, of one month the second time, and similarly subsequently.[78]

The police and administrative authorities soon understood the potential these devices carried. Ultimately, the *module* became an extraordinarily flexible tool that could perform a variety of functions. Thanks to these identification devices, arbitrariness in the surveillance of suspects was formalised and promoted.[79] This kind of surveillance particularly impacted a certain section of urban society. In January 1810, the Vicenza police commissioner explicitly related the *module* to the poverty of many of their owners (the so-called *modulati*):

> The *module IV* and *VII* are issued free of charge since these cards are meant only for suspect people, for those who get out of prison and have been charged with some form of offence, and who are thus mostly miserable and poor, leading the government not to require any taxes or a duty stamp [for the *module*].[80]

The connection between suspicion and social condition (and therefore between social condition and the *module*) was also clear to the local authorities who issued these documents.

Despite the apparent simplicity of the regulations, the unclear nature of some new administrative concepts could generate confusion among police agents. In July 1812, two porters were arrested one hour after sunrise just outside the city walls. Their *module* allowed them to stay within the municipal limits, so they were not committing any felony. However, this was not clear to the police patrol that arrested them: the policemen believed that the city walls served as a border for the whole municipality, whereas they merely delimited the city itself.[81] These police agents were unable to grasp the nature of the new administrative division of the territory – and some *modulati* paid the consequences of this misunderstanding.

Once someone had received a *modula*, it was not easy for them to recover the status of a reliable citizen who could hold an ordinary security card. Theoretically, the *module* did not have an expiry date: receiving a *modula IV* implied that a *modula VII* would be issued once the suspect returned to their hometown, and a *modula VII* could be renewed indefinitely. In fact, it could be renewed simply by being annotated by the police. For instance, a *modula VII* issued in October 1808 was renewed seven times, via handwritten annotations on the back of the card, until July 1809.[82]

It is therefore not surprising that the police archives contain evidence of some attempts to refuse *module*. In May 1808, the police arrested the porter Pietro Fogaresi because he was not carrying the *modula VII* he had received four days previously. In his interrogation, Fogaresi explained: 'last night I came to leave [the *modula*] in this office, [as I was] not happy about being subjected to these measures'.[83] The police had probably gone looking for him as a reaction to this act of resistance. Two years later, a carter and an innkeeper who were suspected of theft refused the *module VII* assigned to them, declaring 'that we do not want accept the aforementioned *modula* for any reason'. They were consequently arrested and imprisoned for three months. This behaviour is testament to the rebellious stance taken by some members of the population against the severe limitations implied by the *module*.

One of the ways to get rid of the status of *modulato* was to send a petition to the authorities. Since the *modula* was assigned by the prefecture, the pleas were normally addressed to the prefect. In December 1812, for example, the prefect of the Bacchiglione received a plea from Carlo Maggiori asking to have his security card returned. In June 1809, he had been arrested in Verona because he was begging even if he was able to work. The police in Verona had deported him to Vicenza with a *modula IV*, where he had obtained a *modula VII*. He had since respected all the provisions stated in the *modula*. Once he had received Maggiori's plea, the prefect asked the police commissioner for advice on the request. Since Maggiori had found a job, had always displayed good behaviour, and had not committed a very serious crime in the first place, the commissioner answered that he deserved to have his *modula VII* withdrawn. Maggiori therefore finally found himself free from police surveillance after three and a half years.[84]

However, these pleas were not always successful. In October 1810, Andrea Novello asked the prefect of Vicenza whether he could obtain a temporary security card. Six months earlier, after seven years of detention for assault, he had been given a *modula VII*. This time the police commissioner gave an unfavourable opinion, because Novello 'does not have means of subsistence and has a suspicious conduct'. The request for a security card was therefore denied.[85] The prefect, however, gave him a *modula IV* valid for two months, which was the validity Novello had requested for his security card.[86] He thus managed to have his *modula VII* temporarily withdrawn, but not to rid himself of police surveillance.

It is difficult to discern a coherent pattern in the decisions the prefect made regarding these pleas for the revocation of *module*. An applicant's good conduct definitely played a part in the decision, but it was not a sufficient factor. For instance, after two years with a *modula VII*, the former counterfeiter Adamo Contariol asked for a security card which would allow him to undertake legal business activities outside the department. Despite receiving a favourable opinion from the police commissioner, the prefect decided that good behaviour alone was not a sufficient factor to strip Contariol of his suspect status.[87] On the other hand, another suspect named Pietro Borò did manage to regain his security card a few months after his arrest, thanks to his good behaviour and, significantly, to the guardianship of his brother.[88] Similarly, the young Alessandro Scoppini had his *modula* withdrawn because the prefecture considered that 'there is no strong reason to keep under a *modula* an individual whose family – with whom he lives – has sufficient means to sustain itself.'[89]

Family sponsorship and guaranteed sources of income were surely positive factors for the authorities. However, other criteria could also play a role in these decisions. In April 1810, the informer Angelo Boezzo regained his security card, thanks to his services to the police.[90] In his case, his ties with the police were a sufficient guarantee of his future behaviour.

As we have seen, state authorities created these cards with the aim of distinguishing reliable citizens from those who were not to be trusted. In this sense, the *module* represented a real-life materialisation of the concept of suspicion. Becoming a *modulato* not only meant being subjected to special surveillance by the police, along with several other limitations to one's freedom of movement; it also carried an implicit social stigma that was recognised and promoted by the state itself. The *module* were at the intersection of the authorities' culture of identification and their will to control "suspect" citizens.

Subaltern Identities, Dangerous Identities

Other traces of the police forces' mentality can be found if we analyse the arrest procedures themselves. Thanks to the arrest reports and interrogation minutes that the Vicenza police commissioner filed between November 1807 and December 1812, it is possible to have a more detailed view of the nature of police control, while reconstructing a social profile of those people who were arrested for irregularities regarding their identity documents. During these five years, the Vicenza police forces arrested at least 248 people for this reason. In most of these cases, the specific motive for the arrest was a lack of identity documents (197), followed by expired security cards (19) and other minor reasons.[91] A large part of these arrests was made by the various police corps (either the municipal police or other forces tied to the police commissioner, 115) and the gendarmerie (80), while very few were carried out by the financial guards (16), the national guard (11), or the army (3).

The arrests themselves were also related to the specific situations in which people were intercepted by the police. As a matter of fact, those whose identity documents were not in order faced varying degrees of danger depending on where they were in the town and on what the time was (Table III).

Table III Times and places of arrest for irregularities in identity documents

When[92]		*Where*	
Morning (7 a.m.–1 p.m.)	16	Public houses	50
Afternoon (1 p.m. –5 p.m.)	10	Streets	47
Evening (5 p.m. –9 p.m.)	38	Inns	40
Night (9 p.m.–7 a.m.)	123	Rural locations	35

The hours of the evening and of the night were the ones in which the police forces conducted most arrests. On the one hand, this shows how people without valid documents were more prone to move at night. On the other, the police forces were certainly more active in their inspection of identity documents during the hours more traditionally associated with crime.

Public places were among the most frequent locations in which arrests were made. This is not surprising, since the Vicenza police commissioner had been exerting a degree of surveillance over these 'suspect places' since 1807. Such places included public houses, inns, and taverns. Unsurprisingly, these places were more likely to be frequented by that mobile, working, and poor population among whom almost all of these arrests were made. However, when not linked to raids on suspect places, document inspections took place randomly on the streets, in both the city and rural areas. In these cases, physical appearance probably played a crucial role in defining who was to be inspected.

Some interesting insights can be drawn from an analysis of the results of these arrests. As we have seen, the concept of suspect was flexible enough to provide the police force with a large margin of manoeuvre. It is therefore not surprising that almost half of these arrests ultimately resulted in the release of the suspect (120). The lack of a valid identity document was indeed a cause for suspicion but not a crime in itself. As the prefect specified in late 1807, when the arrested person could 'justify themselves', then they could be released.[93] In practice, this meant that those who were arrested had to prove their good behaviour in some way. Normally, it was enough to involve a third person, often a family member or an employer. Another possible avenue was to obtain a certificate from a municipality or a parish priest, certifying both the suspect's good behaviour and their conscription status. However, especially for foreigners, it could take some time to obtain the necessary documents. This was the case for the porter Pietro Chiodi, who was arrested for not having a security card in 1810. Chiodi had to wait an entire month in prison for his discharge certificate to come from his municipality of origin, Corteno, a village in the far-away Val Camonica.[94]

Other detainees could also be released, but with limitations to their freedom of movement: 38 were given a *modula VII* and 16, a *modula IV*, or a specifically modified security card that served a similar function. In most of these cases, the people who were arrested were not guilty of any particular crime but had been labelled as suspect by the authorities. It is therefore evident that a large majority of those who had been arrested for not having regular documents could only be charged with being unreliable citizens. Another 50 detainees were entrusted to the prefecture, in most cases, for checks regarding their conscription status. Finally, only 17 of 241 people who were arrested at this time in Vicenza were sent to the courts of justice or the justices of peace because they were actually suspected of being guilty of something. As a matter of fact, the inspection of the regularity of identity documents did not really help to locate criminals, but it was successful in "creating" new suspects.

The police records also allow us to reconstruct the social profile of those who were arrested for a lack of valid identity documents. First of all, only seven women were arrested for irregularities regarding their documents. This small figure can be explained by the fact that women were not seen as subject to the legislation on security cards. The 1802 law that established security cards clearly mentioned that its subjects were 'citizens', but it did not make any gender specification.[95] This vagueness regarding gender requirements mirrored a similar ambiguity in Italian citizenship laws, that did not specify whether a woman could be considered a citizen or not.[96] In practice, however, women did not enjoy the same rights and duties as male citizens.[97] This was also true for the regulations on identity documents. The design of the security cards themselves confirms that they were intended only for men. All the pre-compiled information was printed in masculine form.[98]

However, even this rule could be subverted by the local bureaucracy. In August 1811, the 21-year-old Angela Crescenzia was arrested in Verona for being the concubine of a soldier and, therefore, a suspected illegal prostitute. She was carrying a security card issued by the Vicenza police commissioner at the beginning of the same month. The card had been duly compiled, with just one peculiarity: at the top of the card, '*Il Sig.*' ('Mister'). had been crossed out with a pen. All the other information, including the adjectives in the masculine form, had stayed the same. When she was interrogated by the Vicenza police commissioner, she expressed her surprise at being arrested with a security card: 'I was arrested [...] as a suspect, even if I had the proper security card'.[99]

Not surprisingly, however, the accusation that they did not have the right documents generally puzzled women who were arrested: 'being a woman – one vagrant who was arrested explained – I did not believe that [identity documents] were necessary'.[100] These arrested women often had different types of documents to certify their identities. In May 1810, for instance, a young woman was arrested in an inn under the suspicion of engaging in illegal prostitution. To certify her identity, she carried a certificate of birth issued by her parish priest, together with another certificate issued by a charitable congregation in her birthplace, Treviso.[101] Another young woman – arrested in July 1811 for the same reason – had a birth certificate issued by her parish priest a few months earlier.[102] The fact that security cards were not meant for women did not indicate that women did not certify their identities through other documents.

All of the very few women who were arrested for a lack of valid documents were already suspect for other reason. For instance, the young vagabond Caterina Casagrande was arrested in an inn for being a 'suspect without documents and believed to have been expelled from this department'. As the subsequent inquiry revealed, she had not been expelled before but that she was unemployed and without support: her French "lover" had left her after discovering a few days earlier that she had scabies.[103] Similarly, in March 1812, Angela Comparda was arrested in a rural tavern because she was 'suspect, [a] vagrant and without documents'.[104] In both cases, these women had been arrested as suspects because they had no domicile and displayed a dubious moral and criminal conduct.

Moreover, the relatively small number of arrested women does not mean that women were not arrested in general: the Vicenza police archives contain many reports on arrested women – even if these arrests were probably less frequent than among the male population. Women were simply arrested for other reasons, but not – or not often, at least – for the lack of proper identity documents. It is no accident that, unlike for "ordinary" security cards, women were given *module* quite often. For instance, in June 1810, Anna Ferrarotto received a *modula VII* after six months of detention for theft.[105] Women were even more likely to be given a *modula IV*. In March 1809, three 'wandering gypsies' were deported from the small village of Montegaldella to the departmental capital with a *modula IV*.[106] Similarly, in May 1810, a peasant, Lucia Biganella, arrived in Vicenza with a *modula IV* issued in Padua, where she had been detained for theft for the previous nine months. After withdrawing this first certificate, the Vicenza police commissioner gave Biganella a second *modula IV* to accompany her to her home village.[107] While it was not clear whether a woman could be considered a 'citizen' (and therefore provided with a security card), the local authorities had no doubt that women could be 'suspects' – and thus be given *module*.

Even if the oldest arrested person was 70 years old, the average age of those arrested was approximately 29 (Figure VI).

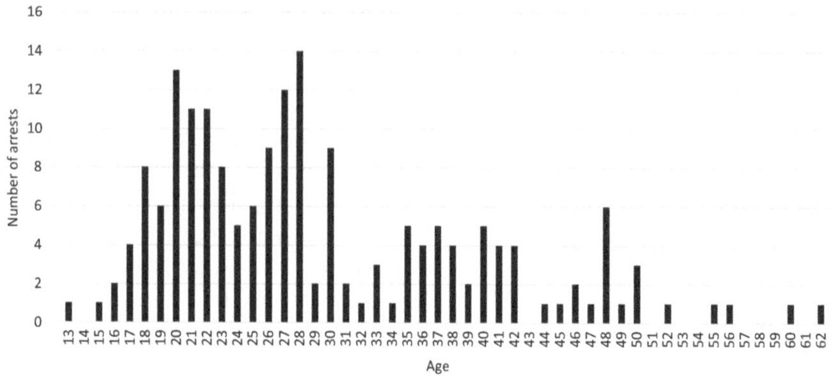

Figure VI Arrests per age.

This fact is even more striking if we consider that, of the 182 people who were arrested whose age is known, 122 were under the age of 30. As we have already mentioned, the departmental authorities explicitly linked the security cards to conscription age.[108] The police forces' focus on young adults was part of a serious effort to limit widespread draft evasion and desertion. To inspect young undocumented people whose age was not certified by a security card was, above all, a way to check on their conscription status.

The decision on how to define the alleged "conscription age" was left to the local authorities and police forces. It is therefore not surprising that several young men under 20 years old were arrested for a lack of proper documents. Some of these teenagers explained that they did not know that it was mandatory for them to carry one of these cards. The cobbler Giuseppe Rizzi, for instance, explained during his interrogation that he did not need a security card, since he was not old enough to be conscripted yet.[109] Similarly, the 19-year-old weaver Angelo Toffanin declared that he did not have a security card because he did not know it was a duty to have one.[110] Even those teenagers who did have a security card could run into trouble during document inspections. In fact, ordinary security cards did not specify the holder's date of birth. It was therefore impossible to know whether a young man had turned 20 before or after he had obtained his security card. This was the case for a young man who was arrested in Vicenza with a security card that had been issued one year previously in Verona, in which he was simply listed as being 19 years old. Although this annotation roused the suspicion of the police, the card itself did not specify whether the young man had already turned 20 or not. Only the subsequent interrogation revealed that he was indeed already 20 years old.[111]

The vagueness surrounding the age at which it was compulsory to carry a security card thus caused confusion both for the general population and for the police forces. This did not, however, lead the authorities to start adding a date of birth to security cards, with the exception of a brief period.[112] The vagueness itself, on the other hand, acted as a precondition for the arbitrary behaviour of the police, who were thus allowed to decide whether or not to arrest young men found without a security card.

Most of the people who were arrested were domiciled in the municipality of Vicenza (117) or in the department of the Bacchiglione (27), while only a minority came from other departments in the Kingdom of Italy (33). Since security cards were originally meant for 'those who travel throughout the territory of the kingdom', this figure could appear surprising. As we will see, urban dwellers tended to use identity documents less often for short trips. However, this also shows that the police forces did not only control the "mobile population" through these identity documents. On the contrary, their arrests primarily targeted the urban population. For a police force that understood itself as external to the local community, every single individual was equally anonymous. Thus, in some way, every single individual was a stranger for the local police. In a city with several thousand inhabitants, this implied that policemen inspected citizens and foreigners with the same level of attention.

However, if police inspections did not vary according to the place of origin of those inspected, the same cannot be said for their social belonging. The higher classes were conspicuously absent among the ranks of those arrested. No landowner, lawyer, or nobleman was arrested in the years considered. A similar remark can be made for members of the Catholic church, who still represented a considerable portion of early nineteenth-century society. Being a wealthy person or a member of the church probably implied a substantial exemption from checks on identity documents. On the contrary, all the professions of the manual working classes were well represented. Rural workers (peasants and shepherds) were the largest category to suffer these arrests. This might be a consequence of the fact security cards were used less frequently among the rural population. The industrial professions, however, were also well represented: weavers, bricklayers, and metalworkers appear often in the files. Several cobblers and butchers were also arrested for irregularities in their documents. Nevertheless, the level of literacy of the arrested is significant. While 111 of them did not sign their interrogation reports, 54 others (31.4 percent) did manage to do so, even if with difficulty.[113]

The above might lead us to see the people who were arrested as mainly belonging to the working class, but, in fact, a large part of them were either unemployed or beggars. During their interrogations, in fact, individuals who were arrested often declared their previous occupation as their actual 'condition'. However, in several of these interrogations, we can glean the actual precariousness of their living conditions. Twenty of the arrested men declared that they were porters (*facchino*), a profession strictly linked to underemployment and unstable incomes.[114] A few dozen other men declared that they were vagabonds, a condition that in most cases implied begging. This information confirms that – along with the working classes – the second social category hardest hit by these arrests was composed of poor people and those who led precarious lives. The social belonging of the arrested men was indeed a good representation of subalternity in Napoleonic Italy.

*

The new state-promoted identity documents found new meanings at the local level. Ever since the Cisalpine period, the central authorities had linked these devices to both securitarian and civil aims. Over the years, this twofold purpose developed into an unavoidable contradiction between a need for surveillance and a will to avoid hindrances to citizens. In the hands of local authorities and the police, the new documents almost completely lost their civil aims. The shortcomings of the issuing authorities themselves rendered security cards unreliable as identification tools. These shortcomings were surely a result of the local bureaucracy's lack of administrative experience. The need to rely on identification procedures which were based on social reputation and relied on unexperienced bureaucrats undermined the new system from the start. However, these deficiencies are also revealing of a widespread lack of understanding of new political concepts, such as the idea of domicile.

If identity documents were unable to provide a means by which to unequivocally identify individuals, the *module* and the police practices related to them offer an example of how the new legislation was twisted in the pursuit of merely repressive aims. Document inspections and decisions on whether or not to assign a *modula* to someone reflect the level of arbitrariness that local authorities adopt in their implementation of the new identification tools. The statistics on arrests for identification irregularities provide a clear indication of how this arbitrariness was directed against a certain sector of society. Inspections of identity documents became, in practice, a way to monitor the poor, working, male and younger sections of the population. As a measure of preventive police, the new identification procedures selectively targeted those categories that had been judged as being prone to delinquency (and draft evasion).

More than creating stable and certain identities, the new identification procedures became, in the urban context of Vicenza, a tool for the surveillance and social repression of those members of the population who would soon come to be defined as *les classes dangereuses*.[115]

Notes

1 In 1809, the Napoleonic authorities estimated 29.918 inhabitants, but they included those who lived in the rural territory of the municipality: *Raccolta delle decisioni, regolamenti, circolari ed istruzioni relative agli attributi proprj delle autorità amministrative nel dipartimento del Bacchiglione. Volume primo* (Vicenza: Da Bartollomeo Paroni. Tipografo Dipartimentale, 1809), 169.

2 On Vicenza during the Napoleonic period: Renato Zironda, ed., *Il vicentino tra rivoluzione giacobina ed età napoleonica. 1797–1813* (Vicenza: Biblioteca civica

Bertoliana, 1989); Emilio Franzina, *Vicenza. Storia di una città. 1404–1866* (Vicenza: Neri Pozza, 1980), 566–623.

3 Andrea Vaccari, *L'attività del commissario di polizia di Vicenza dal 1807 al 1812* (Parma: Università degli Studi di Parma, 1996–1997), 19–20.

4 Theoretically, this legislation had already been extended to the Venetian territories by a royal decree of April 1806: 'Decreto che ordina la pubblicazione, ed esecuzione negli stati veneti di varj decreti in materia amministrativa' (24 April 1806), in *Bollettino*, 1806b, 435–439.

5 Announcement by the prefect of the Bacchiglione (18 October 1806, n. 5389), Bertoliana, Torre, Proclami, vol. A11.

6 'Regolamento per la Polizia generale' (27/93/1802), in *Bollettino*, 1802, 41–47.

7 The 1807 rules established nine different types of security cards meant for different categories of people. However, these types of security cards were not actually put in place before March 1808: notes of the Vicenza police commissioner (3/1808), ASVi, Commissario, f. 2, b. 116. In February 1808, the prefect of the Bacchiglione issued new provisions on the matter in order to 'organise [the issuing of cards] with uniform methods throughout the department'. Only seven types of security cards were mentioned, with different purposes and functions compared with those mentioned in the previous rules: circular from the prefect of the Bacchiglione (5 February 1808, n. 1390), ASVi, Commissariato, b. 1, f. 7.

8 Franco Della Peruta, *Esercito e società nell'Italia napoleonica: Dalla Cisalpina al Regno d'Italia* (Milano: Franco Angeli, 1988), 174.

9 'Regolamento per la istituzione delle Carte di Sicurezza' (undated), ASVi, Commissariato, b. 1, f. 7.

10 Announcement by the prefect of the Bacchiglione (16 July 1807, n. 1826), Bertoliana, Torre, Proclami, vol. A12.

11 Notes of the Vicenza police commissioner (16 December 1807), ASVi, Commissariato, f. 1, b. 65.

12 'Regolamento per la istituzione delle Carte di Sicurezza' (undated), ASVi, Commissariato, b. 1, f. 7.

13 At least in those security cards issued in the same period by the departments of the Adige, Adriatico, Brenta, and Tagliamento that can be found in the Vicenza police commissioner's archives.

14 Security card (n. 1102), ASVi, Commissariato, b. 6, f. 585.

15 The Venetian cards also displayed the serial number of the book from which the card had been taken. An expedient probably made necessary by the great quantity of security cards that the department of the Adriatico had to manage. Security card (n. 1102), ASVi, Commissariato, b. 16, f. 585; security card (n. 224/48692), ASVi, Commissariato, b. 15, f. 358.

16 Interrogation of Bortolo Rocco (12 April 1808); notes of the police commissioner (12 April 1808), ASVi, Commissariato, b. 5, f. 349.

17 Circular from the prefect of the Bacchiglione (30 September 1808, n. 18105), in *Raccolta delle decisioni* 5, 226–227.

18 Announcement by the prefect of the Bacchiglione (18 October 1806, n. 5389), Bertoliana, Torre, Proclami, vol. A11.

19 'Istruzioni ai Vice-Pref.i, Giudicj di Pace e Commassarj di polizia' (undated), ASVi, Commissariato, b. 1, f. 7. The *commissariati di polizia* were established by the new Code of Criminal Procedure in September 1807: 'Codice di Procedura Penale', in *Bollettino*, 1807b, 561–758.

20 Circular from the prefect of the Bacchiglione (5 February 1808, n. 1390); circular from the prefect of the Bacchiglione (1 September 1808, n. 16093), both in b. 1, f. 7.

21 Circular from the prefect of the Bacchiglione (1 October 1808, n. 2948); announcement by the prefect of the Bacchiglione (1 October 1808, n. 2948), both in ASVi, Commissariato, b. 13, f. 1723.

22 With the exception of the three biggest municipalities in the department (Vicenza, Schio and Bassano) where this task was again assigned to the police commissioners.
23 ASVi, Commissariato, b. 15, f. 327.
24 Announcement by the prefect of the Bacchiglione (1 October 1808, n. 2948), ASVi, Commissariato, b. 13, f. 1723.
25 Interrogation of Vincenzo Primon (10 December 1809); notes of the Vicenza police commissioner (11 December 1809), both in ASVi, Commissariato, b. 13, f. 1753.
26 Letter from the Vicenza police commissioner to the prefect of the Bacchiglione (10 May 1810, n. 485), ASVi, Commissariato, b. 15, f. 266.
27 Letter from the prefect of the Bacchiglione to the Vicenza police commissioner (14 May 1810, n. 11583), ASVi, Commissariato, b. 15, f. 266.
28 ASVi, Commissariato, b. 1, f. 63; ASVi, Commissariato, b. 4, f. 208.
29 Interrogation of Angelo Benedetti (30 November 1808), ASVi, Commissariato, b. 9, f. 1073.
30 The 1806 Civil Code stated that a person's domicile was 'the place where [any Italian] has their main dwelling' and that this could change if they moved somewhere else: *Codice civile di Napoleone* 1, 34–36.
31 Circular from the prefect of the Bacchiglione (30 September 1808, n. 18105), in *Raccolta delle decisioni* 5, 226–229.
32 Interrogation of Bortolo Guizzon (1 September 1808), ASVi, Commissario, b. 7, f. 839.
33 The possession of a security card had indeed become the criterion used by the police forces to distinguish between brigands and citizens. Draft circular from the prefect of the Reno to the vice-prefect (16 March 1810, n. 7652), ASBo, Prefettura, 1810, tit. XX, rub. 11 (parte 1), n. 7652/14268.
34 Letter from the police commissioner of the Levante to the prefect of the Reno (8 April 1810, n. 626), ASBo, Prefettura, 1810, tit. XX, rub. 11 (parte 1), n. 14019.
35 Announcement by the prefect of the Bacchiglione (18 October 1806, n. 5389), Bertoliana, Torre, Proclami, vol. A11.
36 Announcement by the prefect of the Bacchiglione (16 July 1807, n. 1826), Bertoliana, Torre, Proclami, vol. A12.
37 Circular from the prefect of the Bacchiglione (5 February 1808, n. 1390), ASVi, Commissariato, b. 1, f. 7.
38 Letter from the prefect of the Bacchiglione to the Vicenza police commissioner (8 September 1809, n. 4629), ASVi, Commissariato, b. 12, f. 1624.
39 Letter from the Vicenza police commissioner to the prefect of the Bacchiglione (14 September 1809), ASVi, Commissariato, b. 12, f. 1264.
40 The cost of a security card, including printing costs and the price of the stamp, varied throughout the years. While in 1807 it was set at 15 cents, the following year, it was raised to 50 cents and finally reduced to 25 cents in October 1809. Security cards were free for those whose inability to work precluded them from getting a job: announcement by the prefect of the Bacchiglione (16 July 1807, n. 1826), Bertoliana, Torre, Proclami, vol. A12; circular from the prefect of the Bacchiglione (30 September 1808, n. 18105), in *Raccolta delle decisioni* 5, 226–229; announcement by the prefect of the Bacchiglione (1 October 1808, n. 2948), ASVi, Commissariato, b. 13, f. 1723. It is not clear when the stamp became mandatory. In fact, a card issued in Verona in January 1806 did not have any stamp: security card (8 January 1806, n. 39), ASVi, Commissariato, b., 5, f. 364.
41 These were, on the one hand, the prefect or the prefecture's general secretary and, on the other, the official actually involved in the procedure.
42 Letter from the prefect of the Bacchiglione to the Vicenza police commissioner (19 April 1808), ASVi, Commissariato, b. 5, f. 372. A note which was

subsequently released by the prefect specified that the holders of these security cards could ask for their cards to be redeemed directly by the employee who committed the inaccuracy: announcement by the prefect of the Bacchiglione (19 April 1808, n. 6448), *ibid.*

43 'Regolamento per la istituzione delle Carte di Sicurezza' (undated), ASVi, Commissariato, b. 1, f. 7.

44 Circular from the prefect of the Bacchiglione (5 February 1808, n. 1390), *ibid.*; circular from the prefect of the Bacchiglione (30 September 1808, n. 18105), in *Raccolta delle decisioni,* vol. 5, 226–232.

45 'Legge per la Polizia' (20 August 1802), in *Bollettino*, 1802, 258–262.

46 'Incumbenze de' commissarj di polizia' (undated), ASVi, Commissariato, b. 1, f. 78.

47 The register contained eleven columns: surname and name, age, birthplace, place of domicile, title of suspicion, period in which the person was declared suspect, injunction, date of injunction, authority of injunction, physical description, and observations.

48 Circular from the prefect of the Bacchiglione (5 February 1808, n. 1390); 'Istruzioni ai Vice-Pref.ti, Giudicj di Pace, e Commisarj di Polizia', both in ASVi, Commissariato, b. 1, f. 7.

49 Circular from the prefect of the Bacchiglione (30 September 1808, n. 18105), in *Raccolta delle decisioni* 5, 226–229.

50 Letter from the prefect of the Bacchiglione to the Vicenza police commissioner (18 December 1807. n. 3394), ASVi, Commissariato, b. 2, f. 84.

51 Letter from Vicenza police commissioner to the Vicenza chief of police guards (23 December 1807, n. 123), *ibid.*; letter from the Vicenza chief of police guards to the Vicenza police commissioner (24 December 1807), ASVi, Commissariato, b. 3, 'Atti riservati', f. 4.

52 'Nota degli oziosi, ladri, sospetti e malviventi, mancanti di modi di sussistenza che presentemente dimorano nella comune e Borghi di Vicenza', *ibid.*

53 'Istruzioni' (undated), ASVi, Commissariato, b. 16, f. 92.

54 Letter from the prefect of the Bacchiglione to the Vicenza police commissioner (17 February 1810, n. 3616/1167), ASVi, Commissariato, b. 14, f. 105.

55 Arrest report (26 April 1812), ASVi, Commissariato, b. 26, f. 266.

56 Interrogation of Bonifacio Comerlato (28 April 1812), *ibid.*

57 The surveillance of foreigners in inns and guesthouses was not an invention of the nineteenth-century authorities. In Venice, for instance, this form surveillance had its origins in the late thirteenth century: Teresa Bernardi and Matteo Pompermaier, "Hospitality and Registration of Foreigners in Early Modern Venice: The Role of Women within Inns and Lodging Houses", *Gender & History* 31/3 (2019): 624–645. Inspections were intensified during the last years of the republic of Venice: Gilles Bertrand, "L'administration vénitienne et l'évolution des techniques d'enregistrement des étrangers dans le contexte de la Révolution francaise (1789–1797)", *Diasporas. Circulations, Migrations, Histoire* 29 (2017): 105–129.

58 Letter from the prefect of the Bacchiglione to the Vicenza police commissioner (24 November 1807, n. 3244), ASVi, Commissariato, b. 1, f. 6.

59 Letter from the Vicenza police commissioner to the prefect of the Bacchiglione (25 November 1807, n. 9), *ibid.*

60 Letter from Giuseppe Amadio to the Vicenza police commissioner (13 June 1811), ASVi, Commissariato, b. 22, f. 452.

61 'Legge normale di polizia', in *Raccolta*, 1/III, 177–189. The law was later mentioned as the legal source underlying the issue of the *modula VII*. However, this Cisalpine law was implemented in the Venetian departments only in June 1808 – almost a year after the establishment of the *module* in the department of

the Bacchiglione: 'Decreto concernente la pubblicazione in alcuni dipartimenti dell'art. 29 della legge normale per la polizia del 10 vendemmiale anno VI', in *Bollettino*, 1808b, 455–456.

62 'Legge per la Polizia' (20 August 1802), in *Bollettino*, 1802, 258–262.

63 'Decreto per l'attivazione della legge di polizia 20 Agosto' (18 October 1802), in *ibid.*, 398–404.

64 Circular from the prefect of the Bacchiglione (30 September 1808, n. 18105), in *Raccolta delle decisioni*, vol. 5, 226–229.

65 Circular from the prefect of the Bacchiglione (1 September 1808, n. 16093), ASVi, Commissariato, b. 1, f. 7.

66 'Regolamento per la istituzione delle Carte di Sicurezza' (undated), *ibid.* Nevertheless, the announcement that actually introduced this regulation only mentioned 'people released from prison' as the intended receivers of this new card: announcement by the prefect of the Bacchiglione (16 July 1807, n. 1826), Bertoliana, Torre, Proclami, vol. A12.

67 'Istruzioni ai Vice-Pref.ti Giudicj di Pace e Commisarj di Polizia' (undated); circular from the prefect of the Bacchiglione (5 February 1808, n. 1390), both in ASVi, Commissariato, b. 1, f. 7.

68 'Regolamento per la istituzione delle Carte di Sicurezza' (undated); circular from the prefect of the Bacchiglione (5 February 1808, n. 1390); Circular from the prefect of the Bacchiglione (1 September 1808, n. 16093), all in *ibid.*

69 After the 1811 reform of the identification system, the documents for suspects were coloured in 'light yellow' to further point out 'the dangerous temper of these individuals': circular from the general police director (12 May 1812, n. 4880), ASBo, Prefettura, 1812, tit. XX, rub. 1 (parte 1).

70 All the *module IV* stored in the archives of the Vicenza police commissioner mentioned their department of origin in the header, except for those from the department of the Mincio, where the header was replaced by a printed 'Prefettura del Mincio'. Moreover, some departments indicated the issuing jurisdiction in the header, while others did not mention it.

71 ASVi, Commissariato, b. 3, f. 176.

72 Interrogation of Giovanni Favero (10 December 1809); letter from the Vicenza police commissioner to the Belluno police commissioner (19 December 1809), both in ASVi, Commissariato, b. 13, f. 1753.

73 *Modula IV* of Giovanni Crevini (21 January 1810); letter from the Vicenza police commissioner to the prefect of the Bacchiglione (14 May 1810), both in ASVi, Commissariato, b. 15, f. 284.

74 *Modula VII* of Andrea Novello (1 March 1810), ASVi, Commissariato, b. 15, f. 295.

75 Letter from the prefect of the Bacchiglione to the Vicenza police commissioner (23 July 1809), ASVi, Commissariato, b. 12, f. 1547.

76 Note by the Vicenza police commissioner (30 December 1811), ASVi, Commissariato, b. 26, f. 252.

77 Nevertheless, Monzerle returned to the inn the following April, where he argued with a costumer and even threatened him with death. He was therefore imprisoned for three days: allocation of a *Modula VII* to Francesco Monzerle (30 December 1808); notes of the Vicenza police commissioner (20 April 1812), both in *ibid.*

78 *Modula VII* of Antonio Boriero (9 June 1812), ASVi, Commissariato, b. 26, f. 333.

79 This arbitrariness also implied that the local authorities could avoid resorting to the *module* for similar purposes. In June 1810, for instance, the Brescian shoemaker Bortolo Pochetto received his expired security card back with an annotation that expelled him from Vicenza: ASVi, Commissariato, b. 16, f. 358. Moreover, the same local authorities could decide not to expel suspects even if

they were not from Vicenza. In May 1808, for example, the baker Pietro Bariani received a *modula VII* after being detained for lack of documents even if he was from Brescia: ASVi, Commissariato, b. 6, f. 483.

80 Letter from the Vicenza police commissioner to the prefect of the Bacchiglione (20 January 1810, n. 92), ASVi, Commissariato, b. 14, f. 45.

81 Interrogations of Francesco Della Vigna and Pietro Lovisetto (28 July 1812), ASVi, Commissariato, b. 28, f. 464.

82 *Modula VII* of Giuseppe Trevelin (24 October 1808), ASVi, Commissariato, b. 14, f. 183.

83 Interrogation of Pietro Fogaresi (28 May 1808), ASVi, Commissariato, b. 6, f. 484. After one month of detention, Fogaresi was released with a new *modula VII* with almost the same provisions: ibid.

84 ASVi, Commissariato, b. 28, f. 820.

85 Letter from the Vicenza police commissioner to the prefect of the Bacchiglione (10 October 1810), ASVi, Commissariato, b. 17, f. 696.

86 ASVi, Commissariato, b. 15, f. 295.

87 ASVi, Commissariato, b. 15, f. 988.

88 ASVi, Commissariato, b. 22, f. 658.

89 Letter from the prefect of the Bacchiglione to the Vicenza police commissioner (26 January 1812, n. 76), ASVi, Commissariato, b. 27, f. 373.

90 Notes of the Vicenza police commissioner (13 April 1810, n. 188), ASVi, Commissariato, b. 19, f. 'Atti riservati'.

91 Violation of *modula VII* (10), forged or stolen security card (10), other irregular documents (4), expired *modula VII* (3) and expired *modula IV*.

92 When interrogated, the arrested people tended to define the hour of their arrest either with a macro-definition (morning, evening, night, etc.) or with a more precise indication. The division of the arrests that I made in this section is therefore approximate and should be considered with caution.

93 'Istruzioni ricercate al Sig.r Cav. Prefetto dal Sig.r Com.o di Polizia di Vicenza' (undated, but 10 December 1807), ASVi, Commissariato, b. 1, f. 78.

94 ASVi, Commissariato, b. 15, f. 329.

95 'Regolamento per la Polizia generale' (27 March 1802), in *Bollettino*, 1802, 41–47. The departmental regulations confirmed this approach: security cards were meant for 'citizens', without further specifications: announcement by the prefect of the Bacchiglione (18 October 1806, n. 5389), Bertoliana, Torre, Proclami, vol. A11; 'Regolamento per la istituzione delle Carte di Sicurezza' (undated), ASVi, Commissariato, b. 1, f. 7. Even when citizenship was not mentioned as a requirement, the authorities maintained a gender-neutral approach. In the announcement linked to the 1807 rules, 'individuals' took the place of 'citizens' as the subjects of the law, with gender specifications being absent in this case too: announcement by the prefect of the Bacchiglione (16 July 1807, n. 1826), Bertoliana, Torre, Proclami, vol. A12.

96 The 1802 citizenship law issued by the Italian Republic stated as follows: 'Anyone who, as of 26 January 1802, has had a stable domicile in the territory of the republic for ten full years, is a citizen'. 'Legge relativa al Titolo II della Costituzione', in *Bollettino*, 1802, 214–220.

97 Nonetheless, even if it is not wrong to consider women as citizens of the Kingdom of Italy, their citizenship was differentiated, as it was in the contemporary French Empire. Rogers Brubaker, *Citizenship and Nationhood in France and Germany* (Cambridge, MA: Harvard University Press, 1992), 43.

98 The empty space for the holder was preceded by the printed formula 'Il' (the male article for "the") and later 'Il Sig'. Moreover, the signature of the holder was preceded by the clear indication 'Firma del suddetto', in the masculine form.

99 Security card and interrogation of Angela Crescenzia, ASVi, Commissariato, b. 23, f. 724.

100 Interrogation of Cattarina Casagrande (26 June 1810), ASVi, Commissariato, b. 15, f. 358. For a similar case, see interrogation of Scolastica Gasparinetti (18 May 1810), ASVi, Commissariato, b. 15, f. 395.
101 Interrogation of Giovanna Conti (26 June 1810), ASVi, Commissariato, b. 15, f. 358.
102 Birth certificate of Lucia Rigatti (19 May 1811), ASVi, Commissariato, b. 22, f. 499.
103 ASVi, Commissariato, b. 15, f. 358.
104 ASVi, Commissariato, b. 26, f. 149.
105 Letter from the prefect of the Bacchiglione to the Vicenza police commissioner (13 June 1810), ASVi, Commissariato, b. 15, f. 332.
106 Letter from the prefect of the Bacchiglione to the Vicenza police commissioner (3 March 1809), ASVi, Commissariato, b. 10, f. 1293.
107 Letter from the prefect of the Bacchiglione to the Vicenza police commissioner (15 May 1810); notes of the Vicenza police commissioner (18 May 1810), both in ASVi, Commissariato, b. 15, f. 279.
108 'Regolamento per la istituzione delle Carte di Sicurezza' (undated), ASVi, Commissariato, b. 1, f. 7.
109 Interrogation of Giuseppe Rizzi (26 March 1812), ASVi, Commissariato, b. 26, f. 200. As a municipal declaration certified, he just had turned 20 – so he had technically not yet been conscripted: *ibid.*
110 Interrogation of Angelo Toffanin (18 April 1808), ASVi, Commissariato, b. 5, f. 368.
111 ASVi, Commissariato, b. 3, f. 166.
112 In 1807, a special type of security card was established for young men who were not yet of conscription age. The card specified that the holder was younger than 20 and had a specific field for the holder's date of birth: 'Regolamento per la istituzione delle Carte di Sicurezza' (undated), ASVi, Commissariato, b. 1, f. 7. However, at the beginning of 1808, this special card was abolished – and young men once again had to certify their identity through ordinary security cards: circular from the prefect of the Bacchiglione (5 February 1808, n. 1390), *ibid.*
113 There are no general statistical data on literacy in Napoleonic Italy. In the department of the Musone, the male literacy rate was 17 percent, although there were significant variations between the city of Macerata (31.2 percent) and the rural municipalities (between 9.2 and 18.8 percent): Elena Brambilla, "Alfabetismo e società nelle Marche in età napoleonica", in *Istruzione, alfabetismo, scrittura. Saggi di storia dell'alfabetizzazione in Italia (sec. XV–XIX)*, eds. Attilio Bartoli Langeli and Xenio Toscani (Milano: FrancoAngeli, 1991), 149–185. In the rural municipalities of the departments of the Mincio the Mella the male literacy rate was higher, even if there were significant variations: Xenio Toscani, "L'alfabetismo nelle campagne dei dipartimenti del Mincio e della Mella e nelle alte valli del Serio e dell'Adda (1806–1810)", in *ibid.*, 201–244.
114 The tendency of unemployed individuals to declare that they were porters was not limited to Vicenza. In Bologna in the 1820s, the most represented social condition between the people subjected to the *precetto politico* was indeed defined as 'facchino': Steven C. Hughes, *Crime, Disorder and the Risorgimento. The Politics of Policing in Bologna* (Cambridge: Cambridge University Press, 1994), 89 (footnote).
115 Louis Chevalier, *Classes laborieuses et classes dangereuses à Paris pendant la première moitié du XIX siècle* (Paris: Plon, 1958).

3 Identity Intermediaries

Civil Registration in the Department of the Adriatico

In late 1810, a young servant by the name of Giovanni Pietro Cotin went to the Venice registry office to publish the banns for his marriage with his co-worker Marianna Panfilio. The civil registrar, an aristocrat named Vettore Sandi, blocked the procedure. He had already done the same with at least two other marriages in the previous year. In his eyes, the problem was clear: even though he was a Catholic, Giovanni was black, while Marianna was white. Sandi believed he could not allow a mixed marriage. As a consequence, the despairing young man considered returning to his homeland of Saint Domingue. A local doctor who visited him diagnosed him with a 'melancholic delirium'. A few weeks later, however, Giovanni made a different resolution. On Christmas Eve 1810, he found a gun, went to the patrician palace where he and his fiancée worked, and fatally shot Marianna. He then tried to kill himself, but only managed to shoot himself in the nose. Marianna was four months pregnant: a desperate attempt was made to save her baby, but the premature child did not survive. In March 1811, Giovanni appeared before the local court. The judge sentenced him to forced labour for life. He was also sentenced to having his right shoulder branded and to one hour of public humiliation.[1]

The Venetian registrar's goal when preventing this interracial marriage was precisely to avoid the birth of mixed-race children. A few months earlier, he had refused to allow another interracial marriage for the same reason. In his view, it was essential that 'from these kinds of unions there should not be born a degenerated, repulsive, and – I would almost say – disgusting race that will propagate for several decades in the following generations'. It was no coincidence that the French Empire had already forbidden interracial marriages in 1804. Even if he was well aware that this provision did not extend to the Kingdom of Italy, the registrar had written to the president of the Venetian civil and criminal court requesting further indications. In the meantime, he had refused to approve any mixed-race marriages.[2] The tragedy that took place on Christmas Eve of 1810 was a direct consequence of his refusal.

The creation of tools to ascertain the legal identities of citizens gave exceptional power to those who oversaw the procedures of identification and registration. A similar power had been wielded, until then, by parish priests.

DOI: 10.4324/9781003412434-4

Among other things, the new civil registrars became crucial actors in defining the gender of new-borns, deciding which marriages could be celebrated, and determining what counted as valid names. This agency was certainly bounded by national regulations. However, the registrars had a free reign in deciding how to apply these national norms and – as in the Venetian case – even temporarily created new rules. But the civil registrars were not the only actors in this bureaucratic game. To operate, the new identification procedures required the participation of several municipal officials, departmental administrators, and members of the judiciary. This chapter will work its way through the maze of these bureaucratic relationships, from the appointment of the registrars to the registers' ultimate use as sources for the definition and certification of personal identities. I will follow the registers themselves, highlighting how these objects were also actors in this complex net of procedures. In so doing, I will narrow my focus to the department of the Adriatico. This department provides an ideal case study on the implementation of civil registration because of its peculiar geographical layout, the presence of a large city (Venice), and its rich variety of archival sources (see Figure III) It was here that a plurality of actors with a plurality of agendas negotiated how the culture of identification that the national authorities had drawn up would be put into practice.

Registration Workers

The registration system envisaged by the *Code Napoléon* placed a municipal official, the civil registrar, at the centre of the registration procedures. It is therefore not surprising that its implementation encountered the same difficulties that the Italian municipal system was experiencing on a general level.[3] According to the 1806 rules, there were two different ways for municipalities to appoint registrars. If a municipality had more than 3,000 inhabitants, the registrar was chosen among the *savi*, a body of officials elected by the municipal Council from among the richest local taxpayers. This was the case of Venice, where the registrar Vettore Sandi was also a municipal *savio*. Most Italian municipalities, however, had fewer than 3,000 inhabitants. In these cases, the mayor was automatically appointed as the civil registrar. Only when the mayor was unable to attend to this task, could the municipal secretary be charged with civil registration procedures. If the secretary did not reside in the municipality itself, the prefect or the vice-prefect could choose a registrar from among the other local municipal officials, following advice from the mayor.[4]

It is not possible to have a precise estimate of how often the task of acting as the municipal registrar was delegated. The delegation procedures were managed directly in conversation between the municipalities and the vice-prefectures. Therefore, the departments did not habitually compile any lists of who the registrars acting within their boundaries were, making difficult – if not impossible – to carry out an overall analysis of the social profile of

Italian registrars. However, it is possible to glean some information on the subject from a specific case.

At the beginning of 1808, the prefect of the Adige drew up a list of all the registrars who had worked in the department over the previous two years. The list contained information on 18 municipalities with fewer than 3,000 inhabitants. Once settled, the position of registrar was quite stable: only four municipalities changed registrars in the two years the list referred to. Over the course of the 2 years, a total of 25 men had held the position of civil registrar in these smaller municipalities. Only 5 of them were mayors, while 12 were municipal secretaries and 8 were private citizens. In most of these municipalities, then, the mayors must have declared that they were unable to act as registrars, but their position was delegated to someone not involved in the municipal administration only in one-third of these cases.[5] It is apparent that the mayors fully exploited the possibility of delegating their registration duties, but that they were also more likely to trust other members of their municipal administration as replacements.

Even if common, the process of delegating registration duties could still be problematic, mainly in cases where there was a lack of alternatives. Two cases in the department of the Adriatico are revealing in this sense. In late 1808, the local justice of the peace in the town of Stra discovered proof of gross misconduct on behalf of the town's registrar, who also happened to be its mayor.[6] Some months later, the latter wrote a long and detailed letter to the prefect of the Adriatico, in which he tried to justify his behaviour. Alongside a series of specific explanations for each of his misdeeds, the mayor mentioned that he had asked to be discharged from his municipal duties at least three times. He largely acknowledged his inability to conduct municipal and civil affairs, and stated that, in his view, this inability justified his mistakes in the civil registration procedures.[7] In one of his attempts to resign from the position of mayor and registrar of his municipality, he had mentioned that his official duties were damaging his professional life – he was a commercial agent, a position that implied a great deal of mobility was thus incompatible with his public duties.[8]

In the same period, a similar stance was adopted by the mayor of Ca' Bianca. In a letter to the vice-prefect of Chioggia, he stated that he and his municipal secretary could not be tasked with carrying out civil registration in the municipality, because they were often out of town.[9] The vice-prefect acknowledged the mayor's good faith and understood that it would have been impossible to appoint him or any other municipal official as the local registrar. He therefore conferred the task to one of the village's inhabitants. He saw this solution as quite uncommon since, in most of the other municipalities in his district, the municipal secretaries acted as registrars.[10]

These two cases show that, in some specific instances, mayors were not in a position to delegate the functions of civil registration. Even if they were more than willing to pass on the task to someone else, it was hard to find citizens who could do the job. The role of registrar surely brought with it more work than benefits. Only in 1808, under pressure from various local authorities, did the Interior ministry finally allow for private citizens who acted as registrars to be paid. However, the ministry categorically ruled out any form of compensation for municipal officials in this position.[11] It is not surprising, then, that mayors tried to delegate civil registration duties whenever this was possible.

However, the choice of who was to act as registrar could also raise issues in municipalities with more than 3,000 inhabitants. In these, as we have seen, the civil registrar was chosen among the town's *savi*, and could rely on the help of one or more municipal employees to carry out the functions related to registration. The *savio* was nonetheless required to attend marriage ceremonies in person – and such events were quite frequent in larger municipalities. In order to make the fulfilment of this unremunerated function sustainable, in 1806, the Venetian *savi* decided to rotate the position of registrar every month. However, after the first month, the provisional departmental authorities asked the appointed registrar, the businessman Angelo Comello, to stay in the position. Comello was shocked: he had only accepted under the guarantee that his duties would not have lasted more than a month. Seriously worried for his private affairs, he proposed assigning the tasks of civil registration to the head of the registry office. A similar solution, he mentioned in his heartfelt letter to the departmental authorities, had been adopted in the municipalities of Padua and Milan.[12] The minister of the Interior denied the registrar's request, pointing out that the legislation provided registrars in the largest municipalities with several forms of aid.[13]

While the dispute in Venice was solved rather quickly, the list of registrars from the department of the Adige, mentioned earlier, suggests that it was not easy, in smaller municipalities, to find a *savio* who was willing to act as registrar for more than a short period. In the municipality of Zevio, for instance, the *savio* who acted as registrar changed at least four times between 1806 and 1807. In other similar municipalities, the position also appears to have been much more unstable than in places where registration duties could be delegated to other municipal officials or private citizens.[14] Senior officials in municipalities with more than 3,000 inhabitants perceived the fulfilment of civil registration duties mainly as a burden on their private lives. The lack of monetary compensation made the position even less appealing. In smaller villages, the position of registrar probably afforded its holders a modicum of symbolic power, but this was less true in larger urban centres. In these, the *savi* were in the difficult – and unpaid – position of having to manage an elaborate bureaucratic procedure imposed from above.

Once the registrars had been appointed, further administrative personnel had to be hired to assist with the civil registration procedures. The legislation allowed for the appointment of 'assistants' to the registrars only in municipalities with more than 3,000 inhabitants. This provision meant that the registrars working in smaller municipalities were left without any form of aid. In May 1806, the government – at the behest of several worried prefects – decided to make it possible for every municipality to appoint adjunct registrars (*aggiunti*), a position already established in France in 1802. Unlike the assistant registrars in larger urban centres, adjunct registrars managed every one of the civil registers – including those for marriages. At the end of the year, they were supposed to pass all their registers on to the municipal registrar. However, local authorities were allowed to appoint adjunct registrars only in those villages that, because of weather conditions, were isolated from their local administrative centres for extended periods of time. In this, the government was thinking specifically of the kingdom's mountainous departments.[15]

Nevertheless, in practice, the position of adjunct registrar was often also used by local authorities in lowland departments to solve the problems posed by the wide territorial extension of several municipalities there. This practice is quite evident in the department of the Adriatico, where the boundaries of several municipalities followed the coast for many miles. By 1812, 24 adjunct registrars had already been appointed in 9 of the 16 municipalities in the district of Portogruaro, and there was still demand for more. The capital of the district itself had no less than six adjunct registrars.[16] Since the district had a total population of about 25,000 inhabitants, this means that the ratio of registration workers (either registrars or adjuncts) to inhabitants was 1 to 625 (see Figure VII).[17] The widespread use of adjunct registrars was also the case in urban municipalities: in 1807, the municipality of Padua employed ten *aggiunti* – almost one for each of the rural villages subjected to the municipality.[18] The entire system would have been unsustainable without the labour of the adjunct registrars.

However, the persistent lack of individuals who were able or willing to manage the civil registers often made the appointment of *aggiunti* difficult. In the department of the Adriatico, the initial appointment of registrars had been quite easy, but it was much more difficult to find auxiliary personnel.

A Matter of Money

Pellestrina, a strip of land 12 kilometres long, straddling the Venetian lagoon and the Adriatic Sea, is a case in point. Its sizeable population of 7,000 inhabitants was concentrated in 3 villages situated at opposite ends of the island: Pellestrina, Portosecco, and San Pietro in Volta. When civil registration was first introduced, Pellestrina was part of the municipality of Chioggia and was thus physically separated from its administrative centre by the sea. Therefore, the local authorities quickly appointed an adjunct registrar for the island. A few months later, however, a delegation of islanders complained to the prefect that the chosen *aggiunto* was morally and physically incapable of managing the local registry. The adjunct registrar was an 80-year-old servant and was unable to cater to such a large population. Every day, he had to register between four and six births, marriages, or deaths. The information which had led the prefecture to appoint the old man, the inhabitants added, was evidently unreliable. They thus asked for the adjunct register to be dismissed,

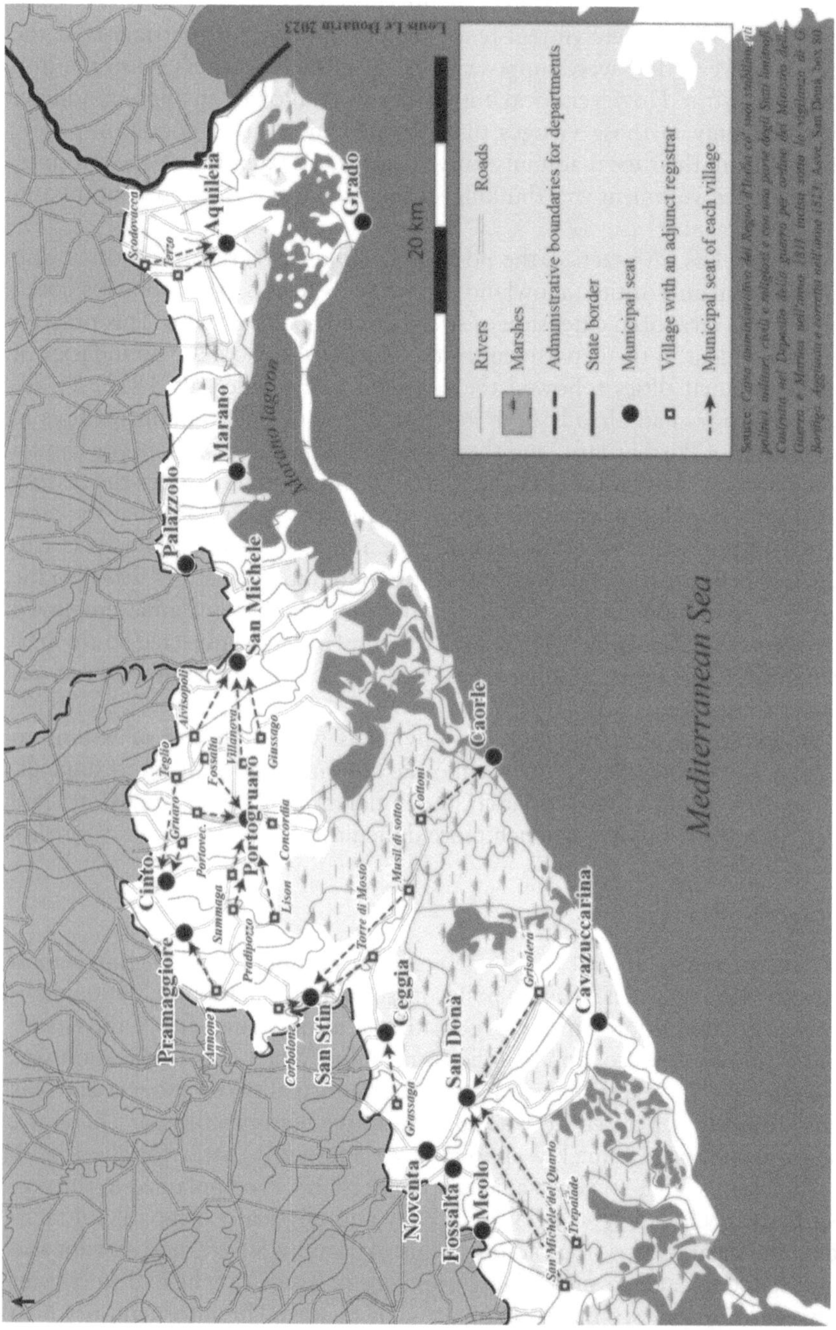

Figure VII Map of the civil registration personnel in the district of Portogruaro in 1812.

and for two new ones to be appointed in his place. They also suggested two suitable candidates.[19]

The subsequent appointment of one of these two candidates as the new adjunct registrar did not solve the problem. This new adjunct registrar wrote to his superior, the registrar in Chioggia, that he accepted the position, but that the size of the island and its large population meant that he needed a paid assistant. He proposed two priests for the job.[20] Even if the registrar in Chioggia supported this request, the first answer from the prefect was negative: the Pellestrina *aggiunto* could enlist the help of local priests if he wanted but was not allowed to officially appoint another adjunct.[21] This solution did not satisfy the adjunct registrar. He wrote to the local representatives in Pellestrina to point out that his profession – he was an apothecary – was not compatible with the management of civil registration.[22] The local judge in Chioggia – who had been temporarily charged with running the municipality – endorsed the request for an assistant, mentioning that the adjunct registrar would otherwise have had to cover large distances alone.[23] The situation was finally solved – after two months of administrative "crisis" – when the prefect decided to appoint a second adjunct registrar for the island.[24] The case of Pellestrina is an excellent example of the practical issues that accompanied the appointment of auxiliary registration personnel. The implementation of the civil registration system often required that local administrative structures undertake a long and arduous process of adaptation to the demographic and social features of the territory in their remit.

Nonetheless, as time passed, the appointment of adjunct registrars proved to be insufficient in solving the structural problems of civil registration in Pellestrina. By 1808, the island had become an independent municipality with its own officials, a mayor and even a justice of the peace.[25] The registrar was now chosen among the municipal *savi* and could benefit from the cooperation of the two *aggiunti* appointed the previous year. This period of stability, however, did not last long. In mid-1808, the island's two adjunct registrars both sent resignation letters to the vice-prefect. They pointed out that they were working for free and were even buying their own office supplies. Moreover, since they were both 'merchants', the position of adjunct registrar negatively impacted their professional lives, while their education level was not high enough to correctly manage the registration procedures. After announcing their resignation, they called for the position to be filled on a rotational basis, since 'every citizen must operate for the good of the population of which he is a part; but this weight must be shared equally, as the sovereign laws prescribe'.[26]

The mayor of Pellestrina, in a subsequent letter to the vice-prefect, agreed, in theory, with the proposal to rotate these unpaid duties. However, he did not approve of the *aggiunti*'s resignation. He pointed out that – far from being 'merchants' – they were shopkeepers, and could thus keep carrying out their registration duties without this having any negative impact on their commercial activity. Moreover, the mayor could not find any alternative adjuncts, since the other literate inhabitants of the municipality were often away from their villages. The mayor's proposal was that, instead of accepting their resignation, the vice-prefect should grant the two adjuncts a salary.[27] The vice-prefect, therefore, rejected their resignation.[28] However, six months later, one of the two adjunct registrars did manage to resign, and the village pharmacist took his position.[29] In the meantime, the introduction of a salary had made the position more appealing than it had been in previous years.

Once they had introduced monetary compensation for those who carried out registration tasks, municipalities encountered new issues. It had previously been difficult to find available individuals for these positions, but the problem now became one of choosing between multiple prospective candidates. When the municipality of Pellestrina needed to find a new assistant registrar in May 1811, conflict erupted between the mayor and the civil registrar. The latter proposed a candidate he trusted, who was in need of regular employment.[30] The mayor, on the other hand, wanted to give the job to a trainee who was carrying out unpaid work for the municipality and also needed a salary.[31] The vice-prefect chose the registrar's candidate, promising that the defeated contender would be considered, as soon as possible, for another paid position.[32] This promise did not reassure the rejected candidate, who left the municipality, causing quite some difficulty for his sponsor, the mayor.[33]

This further fanned the flames of the dispute between the two municipal officials. A few days after the appointment of the new assistant, the Municipal Council was called to ratify the vice-prefect's choice. The mayor took the opportunity to rekindle his conflict with the *savio* who was acting as registrar. He proposed to split the annual salary of the *assistente* (700 Italian lire) with a third municipal official yet to be hired. He justified his proposal by mentioning that the previous assistant had also carried out some other administrative work for the municipality. The *savio*, however, explained that the assistant registrar carried out extra work for the municipality only when he was not busy with his civil registration duties.[34] The Council ultimately sided with the *savio* – and against the mayor – by a large majority. The day after this decision, both the mayor and the registrar wrote to the vice-prefect: the former to contest the town Council's decision, the latter to prevent any further objections from the mayor.[35]

The conflict endured until October 1810. During another meeting of the municipal Council, the mayor tried to argue that the assistant's appointment could not be considered final until the mayor had approved it too. Again, the civil registrar pointed out that the appointment was his own responsibility. This time the Council almost unanimously voted against the mayor's objections: the councillors approved the appointment with only one vote against – probably coming from the mayor himself.[36] When funded, positions in the civil registration system went from being a burdensome obligation to being an economic opportunity that different actors could fight over.

Indeed, the civil registration system generated substantial flows of money at the municipal level. In 1808, the national government estimated that each municipality spent up to 100–110 lire a year, on average, on civil registration procedures. The total annual expenditure for all municipalities in the Kingdom of Italy amounted to around 400,000 lire, a huge sum.[37]

A large part of this sum was used to pay those private citizens who were involved in various ways in the procedures of civil registration. In 1808, 11 municipalities in the department of the Adriatico set aside specific sums of money for their registration personnel: these ranged from 76.5 lire in the small village of Ca Cappello to 1,000 lire in Murano.[38] However, in the civil registration system's first years of operation, municipalities did not have any specific funds with which to cover these costs. Already in 1807, the Legislative Council had discussed the possibility of charging a fee for certificates of births, marriages, and deaths, following the French model.[39] The minister of the Interior opposed the proposal, pointing out that charging fees for these certificates could further discourage people from engaging with the civil registration system, whose usefulness they already found hard to understand.[40] Nevertheless, a solution along these lines was implemented in the following years. Municipalities with more than 3,000 inhabitants were allowed to charge 0.77 lire for certificates of births, marriages, and deaths, along with another fee of 0.38 lire for marriage banns (this included stamp duties).[41]

Even these fees, however, were often not enough to cover the expenses generated by the civil registration system. It is therefore not surprising that, in some cases, local bureaucrats arbitrarily introduced or increased the fees related to certain vital acts and their certificates. In November 1809, for instance, the mayor of Cavarzere reported that the registrar of the adjacent municipality of Cavarzere Sinistro was demanding fees of 1.34 lire, instead of the state-appointed price, for certificates of marriage banns. This disparity in administrative procedures – added the mayor – confused the population of his municipality.[42] Similar cases also emerged in other municipalities in the department. In June 1812, an investigation carried out by the justice of the peace in Aquileia demonstrated that, for years, the local registrar had been demanding exorbitant fees to celebrate civil marriages. The town's mayor had also been aware of this embezzlement: when interrogated, he admitted that he thought that the registrar was just asking for the regular price. This was highly unlikely, since the registrar had been asking up to 4.60 lire – a substantial sum – for each marriage certificate. The registrar, moreover, had continued to ask for extra fees even after the investigation had begun.[43]

Similar practices were also common in smaller municipalities that were not allowed to charge fees for registration procedures. In 1812, two municipalities in the district of Portogruaro – Grado and Murano – admitted that they had received 'small gifts' for issuing certificates and celebrating marriages. The money – the mayor of Grado explained – was used by municipal officials to buy the firewood they needed to keep their office warm in the coldest days of the year.[44] The implementation of the civil registration thus constituted an important financial challenge for small and fragile municipalities too.

The Complex Art of Registering

Once registrars and their auxiliary personnel had been appointed, the next step was to obtain the registers themselves. The 1806 implementation law charged the prefectures and the vice-prefectures with printing and distributing the registers. However, even though the prefectures printed the registers themselves, it was the municipalities who had to cover the printing costs. It is therefore not surprising that some of them – like Murano in 1812 – asked for their civil registers to have fewer pages in order to reduce these costs.[45]

After the registers had been printed, the prefectures were charged with numbering their pages.[46] The prefectures printed the registers with blank pages, even though they made several requests to the government, asking for permission to pre-print the registration forms on the pages of each register. The first of these requests came in May 1806 from the prefect of the department of the Olona. He pointed out that pre-printed registers would have made it easier to implement the registration system in smaller municipalities, where it was almost impossible to find sufficiently literate registrars.[47] The proposal sparked a lively debate both within government ranks and between the central authorities and the prefect of the Olona. After scrutinising imperial administrative practices in nearby Piedmont, the interior minister decided to reject the proposal.[48] Nevertheless, similar suggestions emerged in subsequent years from other local authorities. In July 1808, the municipality of Venice suggested the same solution. To strengthen their proposal, the members of the municipality argued that this practice had already been adopted in Milan.[49] This was not true, but local authorities were happy to use items of fake news to propose administrative practices that could reduce the complexity of the registration procedures for them.

From time to time, local authorities also adopted new procedures not authorised by the national government. In theory, the registers, printed and numbered by the prefectures, had then to be signed by the local courts of first instance.[50] Since these courts sometimes had jurisdiction over large amounts of territory, the work of signing these registers could sometimes take several days. For example, in November 1809, the court of Chioggia needed nine days to sign the 151 registers destined for the territory under its jurisdiction.[51]

Not surprisingly, local authorities came up with different solutions to make this task less demanding. Starting in 1806, the prefecture of the Adriatico charged local justices of peace with the signing of the registers. According to the prefect, this solution reduced the delivery costs for the municipalities, since the justices of the peace were often geographically closer to these than the courts of first instance. The Venetian prosecutor-general, for his part, approved of the practice, as it relieved him and his colleagues at the court of a considerable amount of work. However, even if the court personnel were satisfied with this solution, this was not the case for the justices of the peace. In January 1809, the justice of the peace of Dolo complained that – according to the law – he should not have been tasked with signing the new registers.[52] Despite efforts by the prefect to maintain the practice of delegating the signing of the registers, the interior minister finally put a stop to it in March 1809 – after three years of irregularities.[53]

Local authorities also found other strategies to make these procedures less demanding. In 1809, for instance, the prosecutor-general of Padua began using a stamp to sign the registers, instead of doing so by hand. This went unnoticed until late 1812.[54] Far from the eyes of the central government, local authorities manipulated the regulations on civil registration via a constant process of redefinition and renegotiation.

After this significant organisational effort had been made, the registers were ready to be filled. Each municipality had four types of registers, corresponding to four different legal acts: births, marriage banns, marriages, and deaths. The law explicitly specified that these registers had to be uniform throughout the whole kingdom. The registrar had to compile each of the registers' duplicate sections, leaving no blank spaces between entries and taking care not to leave any loose or empty sheets of paper. What was written in the registry needed to correspond to the declarations of the people involved. It also had to be read out to them after the registry had been compiled. The acts then needed to be signed by all the involved parties, by the witnesses (who had to be male and over 21 years of age), and by the registrar. The regulations contained 11 different example forms that registrars could consult to help them correctly draft each act. Different procedures were also established for the different types of vital acts.

The registration of a new birth had to take place within three days of the birth – and in the presence of the baby if their health allowed it. This registration of birth, which was the duty of the child's father or of any other witness to the delivery, could also take place after baptism. In the case of marriage, the rules were more meticulous. Firstly, the registrar needed to announce the marriage twice, by way of marriage banns, on two consecutive Sundays. Any opposition to the wedding had to be noted in a special register for banns. After the banns, the couple had one year to get married. The regulation (and the Civil Code) precisely described the impediments to, and prerequisites for, marriage, with a special focus on the crucial element of parental consent. This was needed, in one form or another, when the groom was under 25 and the bride under 21 years of age. Civil marriage was a prerequisite for religious marriage and had to be celebrated publicly by the registrar in the presence of four witnesses. As for deaths, their civil registration was needed to proceed with burials. Two witnesses, who, where possible, needed to be neighbours of the deceased or members of their family, were necessary in this case too. The acts also contained a large amount of information regarding who could appeal and who could act as a witness.

The complicated procedures entailed by civil registration provoked a considerable number of mistakes in the compilation of the acts. In 1811, an official from the ministry of Justice analysed all the annual reports that the departmental prosecutors-general had written since 1806 on the management of the civil registers. In his view, 'the inaccuracy, the ignorance and the incompetence' of the registrars had been the first and most important obstacle to the correct use of the registers. He observed that the registrars were not the most educated individuals in their communities – especially in smaller municipalities in the mountains. However, he was confident that a planned merger of smaller municipalities would eventually overcome this obstacle. Moreover, he considered it to be natural that a new system, like civil registration, would require time to be understood by the registrars, especially as far as regarded the 'quite complicated' registration of marriages.

He also listed the 12 departments in which irregularities had been more common: those of the Adda, Agogna, Bacchiglione, Brenta, Lario, Mella, Panaro, Passariano, Piave, Reno, Serio, and Tronto. Considering that there were only 24 departments in total, this suggests that the civil registration system had been inadequately managed in half of the kingdom's territory. Unsurprisingly, almost all of the departments on the list were either partially or mostly covered by mountains. Furthermore, the only lowland department on the list that of the Brenta, was one of the ex-Venetian departments that had been annexed in 1806 and that featured on the government's list of troublesome departments, which included four of the six recently annexed areas. Almost half of the listed department were indeed territories that had only come under French influence relatively recently. Still, the list also included departments that had been created as far back as 1797, during the First Cisalpine Republic, providing proof of how hard it was to implement the civil registration system even in places where attempts to do so had lasted over a decade.

The report also contained a short list of the most common mistakes and irregularities found by the prosecutors-general (Table IV):[55]

Table IV Mistakes and irregularities in the management of the civil registers

Birth acts	*Marriage banns acts*	*Marriage acts*	*Death acts*
• Registration after the legal deadline (three days) • Lack of witness signatures • Lack of the hour of birth • Lack of the day of birth	• Lack of registration • Lack of minutes for the banns • Lack of minutes of the announcement	• Lack of parental or family consent • Lack of ancestors' death certificates • Lack of other preliminary declarations • Lack of witness signatures • Marriages under the legal age • Marriages of widows before the legal term	• Missing minutes • Minutes irregularly written • Lack of witnesses • Lack of witness signatures • Lack of registrar's signature

Most of these mistakes were the result of procedural irregularities. It is worth noticing, for example, that the lack of witnesses is mentioned for three of the four types of acts. Unsurprisingly, the most complicated acts – the ones related to marriage – were the ones that were most often incorrectly compiled. These mistakes were more likely a consequence of the procedures' difficulty than of malice on behalf of the registrars or the general population. Since, in many of the kingdom's municipalities, the civil registration system had to rely on administrative personnel who were not up to the task, it is not so surprising that the more complicated acts were the ones which suffered the most from compilation mistakes. However, the large range of mistakes listed by the prosecutors-general also indicates that the civil registrars often tried to shorten and simplify the difficult procedures prescribed by the regulations. These inaccuracies are testament to the fact that the registrars displayed both an inability and an unwillingness to fully adhere to rules imposed from above.

Faced with this daunting scenario, the government tried to instate a tool that would allow them to keep a constant eye on the quality of the civil registers. It instructed the justices of the peace to inspect the registers of the municipalities under their jurisdiction every two months, and to produce a report based on these inspections for the prosecutors-general. This demanding process gave birth to frequent logistical quarrels between the justices of the peace and the municipalities. The rules stated that the latter were to present their registers to the justices of the peace on a previously agreed date.[56]

On one occasion, the registrar of the small municipality of Contarina missed his meeting with the justice of peace of Loreo, since he was still waiting for registers from two of his adjunct registrars. In cases such as this one, the regulations called for the justice of the peace to personally visit the defaulting registry office, with the trip being paid for by the local municipality. However, the registrar in Contarina decided to send the missing registers by post. The justice of the peace rejected them and travelled to the village the following day to personally inspect the registry office. After his inspection – which brought to light several mistakes and irregularities – he submitted a request for 22 lire to the local municipality, to cover his expenses. The mayor of Contarina, however, refused to pay: in his opinion, the justice's trip was not motivated by the inspection itself but by private businesses.[57] Regardless of the real purposes the justice of the peace had for going to Contarina, this case – similar to dozens of others in the department of the Adriatico alone – demonstrates how local authorities struggled to coordinate the complex logistical procedures entailed by civil registration.

The tangible results of these inspections were the reports sent to the prosecutors-general. The justices of the peace did not follow a uniform template. The justice of the peace of Dolo, for instance, wrote his reports in a narrative style, while his colleague in Asolo preferred a more schematic pattern with sub-sections and columns.[58] The reports could also just be made up of long lists of irregularities. For instance, in January 1808, the justice of the peace of Brendola in the department of the Bacchiglione analysed 60 of the local municipalities' registers after he discovered that they had not been checked since they had first been set up in 1806. The justice of peace – assisted by a scribe and in the presence of the registrar, who stayed silent – listed 142 irregularities in his 15-page report.[59]

But the justices of peace could also prove to be unreliable inspectors. In 1809, for instance, the registrar of the small municipality of Chirignago was accused of committing irregularities in the civil registration procedures. The prefect of the Adriatico thus asked the justice of the peace of Mestre to carry out a special inspection. The latter claimed the accusations were baseless: the registers of Chirignago had been managed perfectly. However, after receiving new accusations, the prefect discovered a blatant conflict of interest: the registrar of Chirignano was also working as a clerk for the justice of the peace in Mestre.[60] The lack of administrative personnel could make it hard to exert a reliable form of control – especially when the people charged with keeping an eye on procedures required surveillance themselves.

Despite the inefficacy and logistical difficulties they entailed, these inspections, nevertheless, remained valid tools through which to invigilate the work of the registrars. The reports were used to locate mistakes and to correct them in good time. In this sense, they primarily served an educational purpose for the registrars, who could learn from their mistakes, while at the same time knowing that their work was being frequently monitored by the judiciary. In a circular to prosecutors and justices of peace issued in June 1810, the justice minister summed up the aim of these inspections: they were useful not just as a way to discover possible mistakes, but also as a way to 'enlighten' those registrars who had a poor awareness of the rules – and to single out those who were completely incapable of managing the civil registration procedures.[61] In the aforementioned internal report on civil registration produced by the Justice ministry, the author argued that well-conducted inspections by the justices of peace could remedy most of the inaccuracy, the ignorance, and the inadequacy of the registrars. However, he acknowledged, these inspections were often conducted in a sloppy manner by the justices.[62]

The government also tried to tackle the problem of the registrars' incompetence at its roots, by making the complex registration procedures easier to understand for those tasked with carrying them out. In 1809, the minister of the Interior published a collection of all the instructions related to the civil registration procedures. The book included the 1806 regulations, along with footnotes containing references to the Civil Code and a number of different circulars on the matter.[63] The collection was the minister's answer to numerous requests for clarification that had come to him from local authorities. He therefore ordered each municipality in the kingdom to buy at least one copy of the collection, of which 5,000 copies had been printed, at a cost of one lira each.[64] In the prefecture of the Adriatico, the vice-prefectures were charged with distributing the booklet and making sure that every municipality had paid for its copy.[65] The collection's importance was such that, during the following year, the minister requested a list of all the copies that had still not been distributed.[66] He then urged every prefect to try to sell their remaining copies. The prefect of the Adriatico, to name one example, published an advertisement for the booklet in the departmental newspaper, possibly in the hope that a few private citizens would buy it.[67]

Similar initiatives came from below too. Also in 1809, a certain Giacomo Giovanetti published a long handbook for civil registrars.[68] The author, a 22-year-old trainee in the judicial administration of Novara, had used a French handbook published three years earlier as the main source for his manuscript.[69] He had then asked the justice minister if he could dedicate the handbook to him. The minister accepted this dedication, on condition that several corrections be made, and this increased the commercial value of the book.[70] The fact that it was dedicated to the minister gave the handbook a sort of semi-official status, making it more credible as a tool that could support the work of the registrars. It is probably thanks to this element of government support that Giovanetti found a publisher in Novara, earning the sizeable sum of 1,151 lire for the publishing rights to the booklet.[71] Considering that the book was sold for 3 lire a copy, the publisher had to sell at least 380 copies only to cover what he had paid for the publishing rights.[72] This considerable investment was probably explained by the handbook's large potential market, since it could be of interest to the thousands of registration workers spread across the Kingdom of Italy.

As a matter of fact, the handbook did indeed circulate widely. In the following months, for instance, the authorities in the departments of the Bacchiglione and the Rubicone asked the national government if they could use it as a guide for their local registrars.[73] On top of its commercial success, the handbook had another concrete consequence for its author. Shortly after its publication, Giovanetti was given the position of employee in the local court in Trento, kick-starting a promising career in the judicial administration.[74] Civil registration, like other expansions of the administrative state, also provided new commercial and career opportunities for ambitious young people.

The Afterlife of Civil Registers

At the end of each year, the registrars had to "close" all their registers with a signature. This act marked the beginning of the registers' second life. Even if they were still annotated in subsequent decades, from that moment on, they became above all a source of information. However, this does not imply that this phase was without difficulty for the actors involved.

During the month of January, one copy of each register was deposited in the town's municipal archives, while the other one was entrusted, along with all its attachments, to the local court of first instance. In practice, however, the delivery of the registers to the courts sometimes did not take place. Both in 1810 and 1812, the Venetian prosecutor-general drew up a list of the registers that were missing from the courts of first instance in his department. The results were discouraging, even if they displayed some signs of improvement over time (Table V).[75]

Table V Number of municipalities in the department of the Adriatico not regularly depositing registers

	1810	*1812*
One year missing	5	12
Two years missing	6	6
Three years missing	11	2
Four years missing	10	/
Total	32	20

Years after the implementation of the civil registration system, the depositing of the registers in court archives was still incomplete. This was true not just for smaller municipalities: in 1810, one of the municipalities of the department of the Adriatico that had not deposited its registers for the previous years was the departmental capital, Venice.

Once the registers had been deposited, the courts of first instance were charged with assessing their condition and sending a yearly report on them to the minister of Justice. According to the implementation laws, the local prosecutor needed to write a notice of inspection in which he had to point out any mistakes or irregularities that could lead to fines or penalties for the registrars. The examination (*verifica*) of the registers could nevertheless raise doubts for the judges charged with the task. In December 1807, the local judge in Chioggia wrote to the prosecutor-general of the Adriatico to ask what *verificare* meant. He pointed out that the French verb *verifier* implied that both the form and the substance of the registers needed to be checked at the same time. The dedicated ministerial circular also specified that the judges had to verify 'the mistakes, the deficiencies, the defects' of the registers. This would have implied an in-depth analysis that would have taken the judge in Chioggia at least ten months: he had 106 registers to examine and was only capable of analysing 80 acts per day. He therefore suggested a lighter interpretation of the term *verifica*: he could limit his examination to those mistakes that had already been found by the justices of peace in their inspections during the year.[76] The justice minister refused this interpretation: he stated that the judges had to examine the substance of the registers and were allowed to take as much time as they needed for this.[77]

When faced with the countless irregularities committed in civil registration, judges could choose whether to punish the registrars or not. The Civil Code prescribed a heavy fine of up to 100 lire for registrars who broke the rules. However, ever since the civil registration system was first put in place, the government adopted a soft line in dealing with mistakes by the registrars. When, in August 1806, the Venetian prosecutor-general asked for permission to fine the registrar in Chioggia – whose registers the local justice of the peace had deemed to be in a foul state – the minister of Justice said no: any fines, he explained, would only be applied after the annual deposit of the registers.[78] In November 1807, the same minister suspended all the fines that had been issued on this matter throughout the Kingdom's territory: before proceeding, he wanted to have a report from each department on the irregularities and mistakes that had been found after the annual deposit of the registers.[79]

The government decided to solve the issue of the fines once and for all in August 1811, with a decree that pardoned all the irregularities committed by the registrars in previous years.[80] When presenting the decree to the viceroy, the minister of Justice acknowledged that the decision not to punish the registrars was due to political expediency, since their mistakes were a result of inexperience and ignorance. Clemency had precedence over severity. Moreover, the possibility of being incarcerated or fined would have made the position of registrar even less appealing, both to municipal officials and to their delegates. The minister therefore proposed to pardon all the previous, incompetence-driven mistakes as part of the measures adopted to celebrate the baptism of the king of Rome.[81] When he sent the decree to the prosecutors-general, the justice minister added a piece of information that was not in the text of the decree itself. From that moment on, the minister wrote, fines and penalties for the registrars were finally going to be applied.[82] After five years, the government's soft approach was both endorsed and terminated at the same time – at least theoretically – with a pardon. One year later, for instance, when faced with the disastrous state of the registers in the small Venetian municipality of Cinto, the authorities decided not to prosecute its former registrars, but rather to enlist their cooperation in the rectification of their mistakes.[83]

The ultimate aim of the civil registration system was to provide legal and definite information on the fundamental vital acts of citizens (births, marriages, deaths), even years after these had taken place. According to the state culture of identification, in the long run, the civil registers were to become the sole source of identity for citizens. Even if the correct creation of the registers implied – as we have seen – a considerable effort, it was only after they had been compiled that they could begin to fulfil their original functions, which were all related to the management of the information they contained. In this sense, the registers provided a public service. Citizens could rely on civil registration to obtain precise information and the legal documents they needed to certify their vital acts.

The laws that implemented the system only allowed for one way of using civil registration for this purpose: requests for certificates. These could be lodged directly with municipal registrars, if the desired act was in the current year's registers. Requests for certificates of acts which had been noted in older, closed registers could only be sent to the courts of first instance. In May 1808, the government interpreted this norm in an extensive way, by also allowing mayors to provide certificates – thus making these requests easier for citizens in rural areas, while also adding to the administrative duties of the municipalities.[84] As has already been mentioned, in larger municipalities, the certificates also constituted a source of revenue, while in the smaller ones, they were issued for free. The only cost for citizens requesting certificates was the stamp duty – which was waived if the applicant had a certificate of property. Thus, the Italian government tried to make the registers as accessible as possible: a stance whose goal was probably to increase the population's familiarity with the novelties of civil registration.

The management of the impressive amount of information contained in the civil registers also generated new technologies.[85] In 1811, the Italian government established alphabetical indexes of all the civil acts registered in every single municipality. This was already a common practice in the French Empire. The origin of this provision was quite incidental, but nonetheless representative of the cooperation that sometimes existed between local authorities and the national government. When the justice minister asked them to send a list of the municipalities in their remit that had not deposited a second copy of their civil registers, members of the Milanese court of justice replied with an unprompted list detailing eight possible ways for improving the civil registration system. This list was forwarded to the interior minister, who noticed the potential usefulness of having alphabetical indexes. Once the justice minister approved of his plan, the interior minister mandated the creation of these indexes.[86,87] Between 1806 and 1814, dozens of local bureaucrats sent similar technical proposals for improving the civil registration system to the national government. The subjects of these proposals varied, but they all shared the goal of facilitating civil registration procedures. Of course, in sending proposals to the central government, these local bureaucrats were probably looking for a way to stand out and advance their careers. But they were mainly trying to make their own lives easier.

The new alphabetical indexes were to be compiled, in duplicate, during the first month of each year. One of the copies would remain in local municipal archives, while the other was to be deposited at the courts of first instance along with the duplicates of the closed registers. Registrars had until the end of the year in which these new instructions were adopted to draw up a first index of all the acts they had registered between the implementation of the civil registration system and the previous year. The instructions also called for the creation of a ten-year index in 1817, which was to be compiled by the courts of first instance and then forwarded to each municipality.[88] The compilation of this first, multi-year index sometimes proved to be more demanding than expected. The municipality of Venice, for instance, asked for, and obtained, a four-month extension of the deadline: even if they had detailed three clerks to the task, by December 1811, they still had to compile indexes for registrations of births and deaths – which were only a part of a total estimate of 74,000 acts.[89] The creation of new tools through which to mobilise administrative information could thus require a substantial initial investment.

Once closed, the registers could also provide the government with a useful source of information for the emerging field of demographic research.[90] In February 1811, the interior minister asked all the prefects in the kingdom to list the precise number of marriages, births, and deaths that had taken place in their departments during the previous year. He asked them to provide data on the departmental capitals in advance, as he was 'particularly' interested in this information. The same request was repeated in subsequent years, and became standard administrative practice.[91] The ministerial circular had been relatively unclear as to how the prefectures were to send this information. Therefore, every department structured the information it sent in a different way. The prefecture of the Adriatico, for instance, only mentioned the number of births, marriages, and deaths, without any further information, and divided these by municipality.[92] The neighbouring prefecture of Passariano transmitted the requested information in a different manner, sending a detailed table in which births were divided according to parents' marital status, marriages according to the marital status of the spouses, and deaths according to the age and gender of the deceased.[93] This difference points to a lack of uniformity in the collection of statistical data that gave prefectural officials space for a certain freedom of initiative.

Once collected by the central government, the departmental reports were assembled in a single national prospectus by the ministry of the Interior. In 1811 and 1812, the national prospectuses were sent directly to the viceroy. A general report included considerations on the differences between the data that had been collected and the statistical averages created in previous decades by demographers. The discrepancies were explained and justified, and the minister also highlighted with satisfaction that the population of the kingdom had increased.[94] Consequently, the viceroy ordered the publication of an article on the subject in the government's official newspaper.[95] The general table was thus published on the *Giornale Italiano* along with a note that pointed out that 'the rising population evidently [demonstrates] how the kingdom [is] prospering'. The note included that this was also thanks to the beneficial effects of smallpox vaccinations.[96] The following year, however, the minister's report to the viceroy was much more concise: for unknown reasons, mortality had increased.[97] This time, the accompanying article on the *Giornale Italiano* was less detailed, mentioning nonetheless that the national population was rising. Above all, it did not contain any open comparison with the data of the previous year, since this would have publicised the rise in mortality.[98] Ultimately, the civil registration system also provided information that the government could use (and manipulate) for its political propaganda.

*

Seen from the point of view of the local authorities, the implementation of civil registration required a herculean effort. A complex set of procedures and practices had to be invented from scratch to turn the identification culture imagined by the Italian government into a tangible and working system. The irruption of the state into certain aspects of the social life of Italian subjects had a cost, and this cost was mostly borne by local officials working on the ground. The civil registration system burdened the weak peripheral structures of the state with a large amount of stress. The municipalities – themselves a recent creation of the administrative state – had immense difficulties in observing all the regulations that were imposed from above. The establishment of municipal registries came with challenges in budgeting, technical aspects, and the deployment of personnel. Higher up, the judiciary were tasked with duties of oversight that compounded the stress caused by their ordinary activities, while the vice-prefectures and prefectures struggled to help municipal officials find the registrars and adjunct registrars they needed. On the one hand, the lack of remuneration made these positions unattractive for municipal officials and private citizens. On the other, when funding was available, conflicts over money erupted both within and between local authorities. This created a situation of constant instability, in which the various actors frequently renegotiated their level of involvement in registration duties.

Nevertheless, over just a few short years, the Italian local authorities were able to make the new civil registration system work across most of the national territory. The effort required by the central authorities was a lot, but it was ultimately bearable. Substantial changes in the administrative division of the kingdom's territory were critical in making this accomplishment possible. However, this success also had a cost – which this time was paid for by the central government. When drawing up rules for the new registration system, the French and Italian authorities had imagined a sort of factory of legal identities. In this "decentralised" factory, everything was to be regulated by detailed rules and mechanisms of control. In practice, however, this plan clashed with the agency of many of the actors involved. Local officials did not behave like disciplined workers. They often did not understand the many complex procedures of the registration system. Even when they did, they tended to simplify them of their own accord, in order to make their duties less burdensome. They fought with each other, they found personal solutions to general problems, and they bombarded the central authorities with suggestions for improvements. Most of all, they registered civil acts in ways that were irregular or flawed according to the rules established by higher authorities.

The success of the civil registration system was paid for by local bureaucrats with their labour and by the central authorities with a degree of tolerance for the former's agency. In this sense, the civil registers were only a partial fulfilment of the original culture of identification dreamed up by the Italian élites. Civil registration did provide a source of legal and stable identities, but the quality of these identities always depended on the various and sometimes conflicting agendas of the several intermediaries involved in transcribing them and fixing them in the registers.

Notes

1 The transcript of the trial is in Giovanni Scarabello and Veronica Gusso, eds., *Processo al moro. Venezia 1811. Razzismo, follia, amore e morte* (Roma: Jouvence, 2000), 92–180. On the case, see: Alfredo Viggiano, "Passions déréglées. Race et justice dans la Venise napoléonienne," in *Récit et justice. France, Italie, Espagne, XIVe-XIXe siècles*, eds. Lucien Faggion and Christophe Régina (Aix-en-Provence: Presses universitaires de Provence, 2020), 243–259.

2 Copy of a letter from the registrar of Venice to the prosecutor-general of Venice (12 July 1810, n. 687), ASMi, Popolazione p.m., b. 54.

3 For the municipal administrations in the Kingdom of Italy, see Pagano, *Enti locali e Stato in Italia sotto Napoleone*.

4 'Regolamento per l'attivazione in tutto lo Stato dei Registri delle nascite, de' matrimonj e delle morti' (27 March 1806), in *Bollettino*, 1806a, 168–223.

5 'Elenco degli Individui che hanno eseguite le funzioni d'Ufficiale, od Aggiunti Ufficiali dello Stato Civile negli anni 1806, 1807 nelle Comuni del Dipartimento dell'Adige' (undated), ASMi, Popolazione p.m., b. 5.

6 Copy of a letter from the justice of the peace of Dolo to the prosecutor-general of Padua (16 November 1808, n. 867), ASVe, Prefettura, b. 226.

7 Letter from the mayor of Stra to the prefect of the department of the Adriatico (22 January 1809, n. 81), *ibid*.

8 Letter from the mayor of Stra to the prefect of the department of the Brenta (undated but circa 09/1807), *ibid*.

9 Letter from the mayor of Ca' Bianca to the vice-prefect of Chioggia (12 January 1809, n. 30), *ibid*.

10 Letter from the vice-prefect of Chioggia to the prefect of the department of the Adriatico (25 February 1809, n. 640), *ibid*.

11 Letter from the interior minister to the general director of the administration of the municipalities (14 November 1808, n. 26488); circular from the general director of the municipalities to the prefects (24 December 1808, n. 7601), both in ASMi, Popolazione p.m., b. 5.

12 Copy of a letter from the municipal Council of Venice to the civil magistrate of the Adriatico (undated but 07/1806, n. 3956), ASMi, Popolazione p.m., b. 3. A few months earlier, the Milanese *savi* had indeed refused to choose a registrar among their ranks, but they had finally accepted it after the intervention of viceroy Eugene: reserved letter from the prefect of the Olona to the interior minister (11 April 1806, n. I/3701); letter from the State secretary to the interior minister (22 April 1806, n. 1206), both in *ibid*.

13 Draft letter from the interior minister to the civil magistrate of the Adriatico (25 April 1806, n. 8175), *ibid*.

14 Each of the nine municipalities with more than three thousand inhabitants in the department had an average of 2.11 registrars in 1806–1807, while the smaller municipalities had an average of 1.38: 'Elenco degli Individui che hanno eseguite le funzioni d'Ufficiale, od Aggiunti Ufficiali dello Stato Civile negli anni 1806, 1807 nelle Comuni del Dipartimento dell'Adige' (undated), ASMi, Popolazione p.m., b. 5.

15 Report from the interior minister to the viceroy (24 April 1806, n. II/3719); copy of the report from the Legislative Council to the viceroy (3 May 1806, n. 55); decree of the viceroy (13 May 1806), all in ASMi, Popolazione p.m., b. 3.

16 See the survey conducted by the vice-prefect of Portogruaro (8/1812), ASVe, San Donà, b. 80.

17 For the approximate population of the district of Portogruaro: decree (22 December 1807), in *Bollettino*, 1807c, 1401–1448.

18 Andrea Desolei, *Istituzioni e archivi a Padova nel periodo napoleonico (1797–1813)* (Cargeghe: Editoriale Documenta, 2012), 459–460.

19 Letter from a local delegation from Pellestrina to the prefect of the Adriatico (24 November 1806, n. 275), ASVe, Prefettura, b. 26.

20 Letter from the adjunct registrar of Pellestrina to the registrar of Chioggia (17 December 1806, n. 1), *ibid*.

21 Draft letter from the prefect of the Adriatico to the registrar of Chioggia (20 December 1806, n. 10482), *ibid*.

22 Copy of a letter from the adjunct registrar to the local representation of Pellestrina (27 December 1806), ASVe, Prefettura, b. 90.

23 Letter from the judge of Chioggia to the prefect of the Adriatico (8 January 1806), *ibid.*

24 The prefect explained that it was not possible to assign an assistant to an adjunct registrar: draft letter from the prefect of the Adriatico to the local representation of Pellestrina (12 January 1807). The suggestion of who the new adjunct registrar should have been came directly from the registrar in Chioggia: letter from the registrar of Chioggia to the prefect of the Adriatico (29 January 1807, n. 24), both in *ibid.*

25 'Decreto che fissa il numero, la residenza, il circondario giurisdizionale, e la classe dei Giudici di pace nei dipartimenti Veneti' (26 June 1807), in *Bollettino*, 1807a, 345–351.

26 Plea from Angelo and Stefano Vianello to the vice-prefect of Chioggia (9 June 1808), ASVe, Prefettura, b. 90.

27 Letter from the mayor of Pellestrina to the vice-prefect of Chioggia (11 June 1808, n. 478), *ibid.*

28 Note by the vice-prefect of Chioggia (13 June 1808, n. 1548), *ibid.*

29 Letter from the mayor of Pellestrina to the vice-prefect of Chioggia (4 December 1808, n. 1042), *ibid.*

30 Letter from the registrar of Pellestrina to the vice-prefect of Chioggia (2 May 1811, n. 37), *ibid.*

31 Letter from the mayor of Pellestrina to the vice-prefect of Chioggia (4 May 1811, n. 516), *ibid.*

32 Draft letter from the vice-prefect of Chioggia to the mayor of Pellestrina (6 May 1811, n. 1774), *ibid.*

33 Letter from the mayor of Pellestrina to the vice-prefect of Chioggia (9 May 1811, n. 550), *ibid.*

34 Minutes of the municipal Council of Pellestrina (13 May 1811), ASVe, Chioggia, b. 39.

35 Letter from the mayor of Pellestrina to the vice-prefect of Chioggia (14 May 1811, n. 570); letter from the registrar of Pellestrina to the vice-prefect of Chioggia (14 May 1811, n. 42), both in *ibid.*

36 Minutes of the Municipal Council of Pellestrina (20 October 1811), *ibid.*

37 Session of the Legislative Council (28 April 1808), ASMi, Registri, vol. 627, 162r–166v.

38 'Spese approvate in via preventiva' (undated), ASMi, Popolazione p.m., b. 5.

39 The proposal, drafted by the commission on municipal administration, envisaged a fee of 0.30 to 0.60 lira for each certificate – while larger municipalities would be able to ask for up to 1 lira for certificates of marriage: draft letter from the Legislative Council's commission for municipal administration to the State councillor in charge of foreign affairs (3 July 1807, n. 59); letter from the Legislative Council's commission for municipal administration the viceroy (13 May 1807, n. 294), both in *ibid.*

40 Report from the interior minister to the viceroy (25 May 1807, n. 7140), *ibid.*

41 *Codice dei podestà e sindaci del Regno d'Italia* (Milano: Dalla Reale Stamperia, 1811), 18.

42 Confidential letter from the Cavarzere police commissioner (27 November 1809, n. 794), ASVe, Chioggia, b. 39. The civil registrar justified this extra fee with a broad interpretation of the ministerial provisions on this regard: letter from the civil registrar of Cavarzere Sinistro to the vice-prefect of Chioggia (9 December 1809), *ibid.*

43 Confidential letter from the justice of the peace of Aquileia to the vice-prefect of Portogruaro (25 June 1812, n. 1219), ASVe, Prefettura, b. 461.

44 Letter from the mayor of Grado to the vice-prefect of Portogruaro (28 April 1812, n. 689); letter from the mayor of Murano to the vice-prefect of Portogruaro (2 May 1812, n. 303); letter from the mayor of Murano to the vice-prefect

of Portogruaro (6 May 1812, n. 314); letter from the mayor of Grado to the vice-prefect of Portogruaro (7 May 1812, n. 809), all in ASVe, San Donà, b. 80.

45 Letter from the vice-prefect of Portogruaro to the prefect of the Adriatico (10 September 1812, n. 8625), ASVe, Prefettura, b. 461.

46 In practice, some prefectures did not carry out the task and passed it on to the courts. In 1811, the justice minister forced the minister of the Interior to issue a circular to the prefects, in order to avoid this practice: letter from the justice minister to the interior minister (1 April 1811, n III/2/7816); draft circular from the interior minister to the prefects (8 April 1811, n. 7808), both in ASMi, Popolazione p.m., b. 7.

47 Letter from the prefect of the Olona to the interior minister (7 May 1806, n. I/3767), ASMi, Popolazione p.m., b. 3.

48 Letter from the interior minister to the justice minister (19 June 1806, n. II/6406), *ibid.*

49 Letter from the prefect of the Adriatico to the justice minister (21 July 1808, n. 14437), ASMi, Popolazione p.m., b. 70.

50 In order to avoid delays, the interior minister set November as the deadline for delivering the registers to the courts of first instance: letter from the interior minister to the justice minister (21 March 1809, n. II/5784), ASMi, Popolazione p.m., b. 6.

51 Draft letter from the vice-prefect of Chioggia to the president of the court of Chioggia (3 November 1809, n. 4044); letter from the president of the court of Chioggia to the vice-prefect of Chioggia (12 November 1809, n. 2166/288), both in ASVe, Chioggia, b. 39.

52 Copy of a letter from the justice of the peace of Dolo to the prosecutor-general of Venice (13 January 1809, n. 965); draft letter from the prefect of the Adriatico to the prosecutor-general of Venice (21 January 1809, n. 1442), both in ASVe, Prefettura, b. 226.

53 Letter from the prefect of the Adriatico to the interior minister (5 March 1809); draft letter from the Interior ministry to the prefect of the Adriatico (11 March 1809, n. 5753), both in ASMi, Popolazione p.m., b. 70.

54 See, for instance, the 1809 civil register of births of Villa del Bosco: ASPa, Stato civile napoleonico, r. 643; letter from the prefect of the Brenta to the interior minister (3 November 1812, n. 26606), ASMi, Popolazione p.m., b. 74.

55 Draft report from the second section of the Justice ministry to the justice minister (18 December 1811), ASMi, Popolazione p.m., b. 8.

56 Circular from the justice minister to the prosecutor of the court of first instance (31 March 1807, n. III/3121/3243), ASMi, Popolazione p.m., b. 5.

57 Letter from the justice of the peace of Loreo to the vice-prefect of Adria (14 March 1810, n. 127); letter from the mayor of Contarina to the vice-prefect of Adria (13 April 1810, n. 126); letter from the justice of the peace of Loreo to the vice-prefect of Adria (18 April 1810, n. 198); letter from the mayor of Contarina to the vice-prefect of Adria (22 April 1810, n. 143); letter from the vice-prefect of Adria to the prefect of the Adriatico (23 April 1810, n. 1517); draft letter from the prefect of the Adriatico to the vice-prefect of Adria (28 April 1810), all in ASVe, Prefettura, b. 286.

58 Copy of a letter from the justice of the peace of Dolo to the prosecutor-general of the Adriatico (16 November 1808, n. 867), ASVe, Prefettura, b. 226; report from the justice of the peace of Asolo on the registers in his canton in the months of November and December 1806, ASMi, Popolazione p.m., b. 85.

59 Minutes of the inspection of the Brendola registers by the justice of the peace of Brendola (7 January 1808), ASMi, Popolazione p.m., b. 73.

60 Draft letter from the prefect of the Adriatico to the general commander of the Police of Venice (12 January 1809, n. 544/687), ASVe, Prefettura, b. 226.

61 Circular from the justice minister to the prosecutors-general, the prosecutors of the courts of first instance and the justices of the peace (9 June 1810, n. II/12399), ASVe, Prefettura, b. 286.
62 Draft report from the second section of the Justice ministry to the justice minister (18 December 1811), ASMi, Popolazione p.m., b. 8.
63 *Raccolta delle istruzioni diramate dal ministero dell'interno del Regno d'Italia per l'esecuzione del decreto di S.A.I. il Principe Vicerè 27 marzo 1806 sui registri degli atti dello stato civile* (Milano: Dalla Stamperia Reale, 1809).
64 Draft circular from the interior minister to the prefects (26 June 1809, n. 1370); invoice of the royal printer with the cost of the collection (10 August 1809), both in ASMi, Popolazione p.m., b. 6.
65 Draft circular from the prefecture of the Adriatico to the vice-prefects (24 July 1809, n. 12569), ASVe, Prefettura, b. 207.
66 Letter from the director of the royal print-shop to the interior minister (7 May 1810, n. 922), ASMi, Popolazione p.m., b. 6.
67 Letter from the interior minister to the prefect of the Adriatico (2 May 1810, n. 4953); draft letter from the prefect of the Adriatico to the interior minister (8 May 1810, n. 8910), both in ASVe, Prefettura, b. 286.
68 Giacomo Giovanetti, *Manuale degli ufficiali dello stato civile in cui sono gradatamente tracciate le operazioni da farsi nella formazione de' singoli atti e le module per ciascun d'essi* (Novara: dalla stamperia di Giuseppe Rosario, 1809).
69 Giovanetti changed and added some new material to the original text: M. Charvillhac, *Le guide de l'officier de l'état civil, ou Méthode simple et facile pour diriger la marche qu'il doit suivre dans la rédaction des actes de son ministère* (Paris: Chez Rondonneau, 1806).
70 The internal report on Giovanetti's manuscript highlighted several inaccuracies but praised the author's 'good will and propensity to work': report to the justice minister by Borella (undated), ASMi, Popolazione p.m., b. 6.
71 Publishing contract (7 August 1809), Archivio di Stato di Novara, Giacomo Giovanetti, b. 22.
72 *Giornale italiano*, 19 January 1810; *Il corriere milanese*, 31 January 1810.
73 The national authorities acknowledged the usefulness of the handbook, but clarified that it could not replace the official regulations: draft letter from the director of the municipal administrations to the prefect of the Bacchiglione (23 March 1810, n. 1761), ASMi, Popolazione p.m., b. 73. Letter from the prefect of the Rubicone to the interior minister (24 April 1810, n. I/7451/2174); letter from the interior minister to the justice minister (28 April 1810, n. 8390); letter from the justice minister to the interior minister (29 April 1810, n. II/8899), all in ASMi, Popolazione p.m., b. 6
74 Giovanetti's career ended with the end of the Napoleonic government in 1814. He nevertheless became an influent intellectual and politician in the Restoration Kingdom of Sardinia: Franco Della Peruta, "Giovanetti, Giacomo", in *Dizionario Biografico degli Italiani*, vol. 55 (2001): 447–450; Elisabetta Fiocchi Malaspina, *L'utile giusto. Il binomio economia e diritto per l'avvocato Giacomo Giovanetti (1787–1849)* (Soveria Mannelli: Rubbettino, 2020).
75 Copy of the list of municipalities that had not delivered the closed registers to the courts of first instance in the department of the Adriatico (undated, but 05/1810), ASVe, Prefettura, b. 286; list of municipalities that had not delivered the closed registers to the courts of first instance in the department of the Adriatico (5 March 1812), ASVe, Prefettura, b. 461. These lists were the result of a new policy enacted by the minister of Justice in 1810 to check on the actual delivery of these registers. He had ordered the prosecutors-general throughout the kingdom to verify each year whether the registers were delivered or not, since this duty had

been neglected by several registrars: letter from the justice minister to the interior minister (10 April 1810, n. II/7380–7251), ASMi, Popolazione p.m., b. 6.

76 Copy of a letter from the Chioggia prosecutor to the prosecutor-general of the Adriatico (8 December 1807), ASMi, Popolazione p.m., b. 70.

77 Draft letter from the justice minister to the prosecutor-general of the Adriatico (4 January 1808), ASMi, Popolazione p.m., b. 70.

78 Letter from the prosecutor of Venice to the justice minister (27 August 2020); draft letter from the minister of Justice to the prosecutor of Venice (30 September 2020, n. III/8558), both in *ibid.*

79 The circular is quoted in a draft circular from the justice minister to the prosecutors (30 January 1808, n. 2372), ASMi, Popolazione p.m., b. 5.

80 The decree was issued on 27 August 1811. The preparatory work and the final text of the decree can be found in *ibid.*

81 Draft letter from the justice minister to the viceroy (undated, n. II/12072), ASMi, Popolazione p.m., b. 7.

82 Circular from the justice minister to the prosecutors and the justices of the peace (8 September 1811, n. 20219), *ibid.*

83 Letter from the mayor of Cinto to the vice-prefect of Portogruaro (3 April 1812, n.383), ASVe, San Donà, b. 80.

84 Circular (17 May 1808, n. 11455), quoted in *Raccolta delle istruzioni diramate*, 18 (footnote). The same year the government also intervened to specify that the *estratti* needed to be a perfect copy of the original acts: Draft circular from the Interior minister to the prefects (13 June 1808, n. 12938), ASMi, Popolazione p.m., b. 5.

85 Denis and Lacour, "La logistique des savoirs", 107–122.

86 List of the municipalities that had not deposited the 1810 register in the department of the Olona (undated, but 03/1811); letter from the justice minister to the interior minister (6 March 1811, n. II/3190); draft letter from the interior minister to the justice minister (12 March 1811, n. 5685); letter from the justice minister to the interior minister (21 May 1811, n. 5920), all in ASMi, Popolazione p.m., b. 7. Actually, the French decree was mainly concerned with the ten-year indexes: *La Législation civile, commerciale et criminelle de la France ou commentaire et complément des codes francais [...] par m. le baron Locré, Tome Troisième* (Paris: Treuttel et Wurtz, 1827), 280–282.

87 Letter from the general director of the municipal administrations to the interior minister (18 June 1811, n. 4916); draft circular from the interior minister to the prefects (24 July 1811, n. 17173), ASMi, Popolazione p.m., b. 7.

88 Instructions written by the interior and justice ministers for the alphabetical indexes of civil registers (undated), *ibid.*

89 Letter from the mayor of Venice to the prefect of the Adriatico (18 December 1811, n. 14036); draft letter from the interior minister to the prefect of the Adriatico (8 January 1812, n. II/636), both in ASMi, Popolazione p.m., b. 70.

90 On official statistical research in the Kingdom of Italy, see Terenzio Maccabelli, "Melchiorre Gioia e la Statistica del dipartimento del Mella", *Commentari dell'Ateneo di Brescia* 399 (2005): 399–432.

91 Draft circular from the interior minister to the prefects (16 February 1811, n. 3984); draft circular from the interior minister to the prefects (17 December 1811, n. II/29438), both in ASMi, Popolazione p.m., b. 7; circular from the interior minister to the prefects (16 February 1813, n. 3984) mentioned in the letter from the prefect of the Piave to the interior minister (2 February 1813, n. 1269), ASMi, Popolazione p.m., b. 80.

92 Draft letter from the prefect of the Adriatico to the interior minister (2 May 1811, n. 8244), ASVe, Prefettura, b. 354.

93 Table attached to the letter from the prefect of the Passariano to the interior minister (16 April 1811, n. 8814), *ibid.*
94 Report from the interior minister to the viceroy (11 July 1811, n. II/16275), ASMi, Popolazione p.m., b. 7.
95 Letter from the state secretary in Milan to the interior minister (20 July 1811, n. 3965), *ibid.*
96 *Giornale Italiano*, 28 July 1811, 836. On smallpox vaccination: Alexander Grab, "Smallpox vaccination in Napoleonic Italy (1800–1814)", *Napoleonica. La Revue* 30 (2017): 38–58.
97 Draft report from the interior minister to the viceroy (15 July 1812, n. II/20067), ASMi, Popolazione p.m., b. 7. Consequently, the viceroy asked for further explanations on the rise of mortality: draft letter from the state secretary to the interior minister (11 September 1812, n. 3000), ASMi, Popolazione p.m., b. 8.
98 *Giornale Italiano*, 16 July 1812, 794.

4 Ambiguous Partners

The Clergy

In December 1808, the Italian government enlisted the help of the clergy to explain the importance of civil registration to the general population. As a first step, the minister of Worship, Bovara, asked the bishops to explain to the hundreds of parish priests scattered across the kingdom exactly *why* the civil registers were so important. The bishops were given a free hand to draw up their own arguments on the subject, as long as the message was transmitted.[1] Twenty days later, the bishop of Udine, Baldassare Rasponi, sent a four-page circular to all the parish priests in his diocese. In his elaborated text, Rasponi argued that a Christian submission to god implied a complete submission to those political powers 'to whom god himself has given the task of governing the earth'. Rasponi claimed that this reason alone would have been enough to convince priests to faithfully observe the new registration duties imposed by the Napoleonic government. However, the bishop also took his arguments a step further. Rasponi fervently exhorted his priests to 'multiply our prayers for [Napoleon], and – at the same time – to recommend his valorous armies to god'.[2] Despite the bishop's prayers, however, it was the Habsburg armies who entered Udine a few months later, in April 1809. Faced with this abrupt change of government, the bishop – who had recently been flaunting his Napoleonic credentials – instructed the clergy to perform the traditional *Te Deum* in the cathedral to celebrate the return of Udine's former rulers. Then, after just a few weeks, Napoleon's army returned to Udine, and Rasponi was sentenced to death by the emperor himself, outraged by the bishop's fickle loyalty.[3]

Even if Rasponi managed to have his sentence commuted to indefinite internal exile, his case offers a good representation of the ambiguous relationship that the Italian clergy maintained with Napoleonic governments. Recent work by historians has moved beyond the idea that the interactions between the Italian clergy and Napoleonic authorities can be described only as a culture war.[4] In the case of civil registration, it seems apparent that this relationship was much more layered and nuanced – and even contradictory. As Rasponi's case shows, the Italian clergy did cooperate with Napoleonic governments in their attempts at state building. However, this cooperation was based on ambiguous premises and could

DOI: 10.4324/9781003412434-5

quickly fail. As we have seen in Chapter 1, the Italian élites were animated by secularist stances. Their interventionism in religious matters provoked forms of discontent in the Italian clergy, especially after the annexation of the Papal States to the Empire and the imprisonment of pope Pius VII in 1809.[5] However, in the Kingdom of Italy, the clergy never interrupted its cooperation with the Napoleonic authorities. Neither did it engage in forms of open resistance such as, for instance, the refusal by almost half of the priests in the French department of Rome to swear allegiance to the government.[6]

The identification and registration duties of the clergy take us to the heart of the ambiguous relationship between the Catholic Church and the Napoleonic states. The clergy, at every level, engaged with older and newer practices aimed at identifying Italian citizens. Its culture of identification and its effective approval of the government were constantly questioned by the growing intervention of the Napoleonic state in this field. Nevertheless, bishops, priests, and chaplains were crucial actors in the definition of Italian identities at the beginning of the nineteenth century.

Old Religious Procedures, New State Interferences

While the civil authorities were working on their new systems of registration and identification, the clergy continued to manage the traditional ones. The parish priests' pastoral role in their communities gave them detailed knowledge of their parishioners. This knowledge found its embodiment in the parish family books (*stati d'anime*) that priests had been compiling in Catholic Europe since the beginning of the seventeenth century.[7] These books were traditionally compiled by parish priests each year at Easter. They contained information both on the family and personal status of parishioners and on their religious practices.[8] Thanks to these books, each year, the authorities of nineteenth-century Italian states gathered demographical statistics on their population.[9] However, after the introduction of civil registration, the Cisalpine government considered this type of census to be an outdated practice, and even considered terminating it.[10]

Even if the Napoleonic government started to collect similar data from its civil registers, the local clergy continued to compile these parish family books. The civil authorities also found themselves resorting to this older form of parish-based knowledge several times at the beginning of the nineteenth century. The police law of 1802, for instance, charged priests with compiling lists of all those inhabitants in their parishes who did not have a permanent residence.[11] In 1812, when the prefect of the Musone started an inquiry on the individuals registered without a surname in his department, he used the parish family books as a source to correctly identify those citizens whose surname was missing.[12] Moreover, a few months earlier, the parish priests had helped to compile general population lists in several municipalities. In Venice, for instance, they had assisted the civil officials charged with the door-to-door registration of the city's inhabitants.[13] The civil authorities recognised, and often resorted to, the priests' longstanding expertise and knowledge of local communities.

Even after the introduction of civil registration, parish priests continued to manage the religious registers of births, marriages, and deaths. As we have seen in Chapter 1, state authorities had recognised right at the beginning of the nineteenth century that these parish registers were not easily replaceable as public sources of information. As a consequence, they began trying to extend their jurisdiction over a matter that, in the past, had been largely left in the clergy's own hands. This new interest from the government carried with it new duties for the parish priests. In 1802, for instance, the interior minister began requesting monthly copies of the parish registers, adding to the work of priests who, in some dioceses, already had to send an annual copy of their registers to the local bishop.[14] Moreover, any rectifications to closed parish registers were now subject to the authorisation not just of the local diocesan curia, but also of the civil authorities. In June 1808, for example, the vice-prefect of Chioggia discovered that the local bishop had accepted a plea from a man who wanted to correct his baptismal registration. The vice-prefect promptly wrote to the bishop to ask about the 'the legal proof of the necessity of this correction'.[15] He then explained that the correction could only be authorised if it was properly justified and only – of course – if he himself approved.[16] The clergy was still in charge of these closed parish registers, but was subject to increasing surveillance on behalf of the civil authorities.

Something similar also happened with the traditional parish certificates, the *fedi*. These were signed pieces of paper that certified information taken from parish registers of births, baptisms, marriages, and deaths. Parish priests could also issue certificates on the good behaviour of their parishioners, also known as *fedi*. The text and layout of these *fedi* varied wildly, demonstrating a high degree of flexibility and deregulation. The state authorities left the *fedi* largely unregulated, but they ultimately acknowledged their public value. In 1805, the Italian government made it mandatory to use stamped paper for the *fedi*. In this way, it strengthened the official and legitimate nature of the certificates.[17] The parish *fedi* could also be substantiated by local authorities with a declaration of veracity signed by the municipality and a state-approved stamp.[18] In 1807, the government even made authorisation by the civil authorities mandatory for parish certificates issued to conscripts.[19] In 1811, the equivalence between civil and religious certificates was further strengthened, when a decree declared that certificates issued by civil registries would be exempt from stamp duty.[20] After several local prosecutors asked for clarifications, the minister of Justice gave an extensive interpretation of the decree. He argued that parish certificates should also be exempt from stamp duty, 'considering that the priests, in issuing these certificates, do substantially the same [work] as the civil registrars'.[21] The management of religious registration was therefore equated to the management of its civil counterpart, and the role of the parish priests was implicitly equalised to that of the civil registrars.

The production of these certificates thus remained in the hands of the parish priests even after the establishment of state identity documents. These parish certificates actually remained necessary for the fulfilment of a vast number of social acts. As we will see in Chapter 5, the *fedi* continued to be the most commonly used identity documents in the Kingdom of Italy. Any individual who travelled a certain distance from their usual place of residence would have, in most cases, brought with them at least one birth certificate along with their state identity documents.

Parish certificates were also necessary for other purposes. In late 1798, for instance, a house painter from Varese wrote to the Cisalpine Executive Directory because he could not obtain the birth certificate of his deceased uncle from the local parish priest. He needed this *fede* to be able to claim the considerable sum he stood to inherit from this uncle, who had died without heirs two years earlier in Paris.[22]

At the same time, the state authorities' increasing interventionism in the everyday lives of Italians led to a greater demand for parish certificates. Citizens had to present their *fede* of birth, for example, to obtain civil certificates of good behaviour.[23] Parish certificates were also needed as a prerequisite for the celebration of civil marriages. Both bride and groom had to present their birth certificates before the ceremony, but they also – when under 25 if male, and under 21 if female – had to prove any eventual deaths among their parents and grandparents with specific *fedi*.[24] The use of parish certificates as sources for civil records was thus a standard practice. When a general population list was first drawn up for Milan, for instance, its inhabitants had to justify their date of birth with parish certificates.[25] The demand for *fedi* could sometimes be so high that parish priests had to stop issuing any other certificates. This was the case in Cannobio when the conscription lists for 1806 were being compiled. The use of pre-printed forms in this case did not accelerate procedures either.[26] The state's goal of building an autonomous system of registration and identification rested on an extensive use of parish certificates – and significantly increased the involvement of parish priests in its procedures.

Since the *fedi* continued to be crucial in the everyday life of Italians, from time to time, the clergy still had to intervene to correct mistakes or shortcomings in the old parish registers. As already mentioned, the civil authorities had already begun to intervene in these procedures. However, it was mainly the dioceses who were responsible for amendments to old parish registers. If something had been noted incorrectly in a parish register, the involved parties usually submitted a plea to the diocesan authorities.[27] In July 1811, for instance, Anna Maria Dalla Presepia asked the general vicar of Vicenza if her parish birth registration could be corrected. To her surprise, when she had asked the parish priest of the village she was born in for a *fede* based on her birth registration, the latter had issued a certificate in which her father's name was different from the one she knew. As she later discovered, in the 1770s, her father had used different names in different situations since 'he had fallen out of favour with the former Venetian government'. In order to corroborate this information, she attached eight different parish certificates to her plea – thus ensuring a quick approval of her request, with no further investigations.[28]

The provision of written evidence, however, was not the only way to have a parish register rectified. In February 1811, a couple wrote to the diocese of Vicenza because their sons' christenings had not been registered more than 35 years earlier. They needed their sons' birth certificates for unspecified 'obligations of the law'. Since they had not attached any written evidence to their plea, the general vicar opened an investigation. Four people were summoned and questioned: two of the children's godparents, a neighbour, and one of the couple's servants. Even if they did not remember the precise dates of the births, all the witnesses certified that the couple's sons had been born in the months and years declared by the couple in their plea. One of those interrogated confirmed this information by relaying a joke she remembered hearing at the time: since both children were born in October, the couple's friends used to say that 'Cecchetta [the mother's nickname] always gives birth in October, when the good wine comes'. After the interrogations, the couple managed to have their sons' baptisms registered.[29] At the same time in which the state was building its civil database of its citizens' social acts, the clergy continued to manage its own one, conducting investigations and inquiries to verify "real" identities and registrations.

The old parish registers were not the only means by which Italian clergymen continued to hold sway on the registration system. Older interpretations of the sacraments often persisted even after the extension of civil jurisdiction to marriages and baptisms. On several occasions, members of the clergy appealed to these interpretations to request changes to civil laws that would preserve the traditional practices allowed by canon law. In 1803, for instance, a parish priest in Lodi asked to keep the birth of a newborn secret. Since the child had been born nine-and-a-half months after the death of her father, the clergyman had christened her as the daughter of an unknown father. He had celebrated a secret baptism, in order to save the mother from a potential scandal – he exhorted the municipal registrar to follow his lead, and keep the child's civil registration secret too. Despite the priest's plea, the registrar decided the child was legitimate after receiving assurances from medical experts that a pregnancy could last even longer than ten months.[30]

Secrecy was also at the centre of another request from the clergy, made by the vicar of Verona in June 1806. In the previous months, two different couples in his remit had asked to be married in secret. In both cases, the applicants were publicly living as married couples even if they had never legally wed. Both couples had now repented and wanted to regulate their condition without causing a scandal. The local registrar, however, had not allowed the secret marriages to take place. The vicar thus asked if a civil register for secret marriages could be instituted, since this had already happened in Verona in 1803, even if the Civil Code had made the practice illegal a few months later. As an alternative, he asked if at least secret religious marriages could be allowed, even without civil effects, so as to solve the problem. As he rhetorically pointed out

> The civil laws do not forbid fornication: why on earth should the same laws forbid the Christian secret marriage that is necessary to appease one's conscience and to be preserved from dishonour or serious damage?[31]

Issues around marriage often revealed the persistence of traditional understandings of the sacrament among clergymen. A few months after the episode in Verona, the parish priest of Vidigulfo asked the authorities of the department of the Olona to annul a civil marriage that had been celebrated in his parish. According to the priest, the groom was impotent: this was well known in town and had been confirmed by a medical inspection. He wanted the marriage annulled because it had already provoked 'grave scandal', and because he could not celebrate a religious marriage if the groom was impotent. Also in this case, the authorities refused the priest's proposal and stuck to the dictates of the Civil Code: according to the civil law, and contrary to the canon law, impotence was not a valid impediment to marriage.[32] The religious understandings of sacraments inevitably entered into conflict with the principles introduced by civil legislation.

It is not thus surprising that, especially in the first years after the Civil Code entered into effect, Italian clergymen gave religious procedures precedence over civil ones. This was the case, for instance, for civil marriages whose religious counterparts could not be celebrated because of canonical impediments. As stated by the 1803 Concordat, state authorities could not force parish priests to celebrate marriages that were against canon law.[33] In these cases, couples who required a civil marriage were forced to request a matrimonial dispensation from the local diocese – a procedure that entailed a substantial cost.[34]

Moreover, in the early days after the implementation of the Civil Code, parish priests did not immediately begin to ask for proof of the civil registration of a marriage before performing religious marriage ceremonies, as prescribed by the Code. In January 1807, for instance, the registrar of Chioggia learned that a religious marriage had been celebrated for two members of prominent families from Pellestrina, an island under his jurisdiction. A subsequent investigation showed that the local parish priest had already celebrated seven marriages that had not been inscribed in the civil register – which had been in operation in Pellestrina since December 1806. When the situation emerged, the priest tried to justify himself by blaming the registrar in Chioggia, who had apparently not informed him of his new duties. He said that he did not know that marriages had to be inscribed in the civil register before they could be celebrated in church.[35] Despite his justification, the priest was reprimanded by the vicar of Chioggia, who also punished him with eight days' penance in a local monastery.[36] It thus took some time for the interweaving of civil and religious procedures to be fully understood – and observed – by parish priests.

State interventionism in religious registration procedures provoked a variety of responses among members of the clergy. After the introduction of the Code Napoleon, the Italian government intervened in christening procedures on at least two different occasions. The first of these two occurred when the central health authority of the department of the Olona specified that parish priests in the department were to use only warm water when baptising newborns in winter. According to the health authority, the use of cold water was provoking 'serious harm' to several newborns.[37] A few parish priests disdainfully replied that they were already using warm water.[38]

The health of newborns was also at the centre of the government's second intervention in baptism procedures. In January 1807, the prefect of the mountainous department of the Piave realised that very few of the young men registered in the department's conscription lists had been born in the winter. A subsequent inquiry found that the department had a shocking infant mortality rate of 80 percent, which, according to the prefecture, was largely due to the fact that infants were transported across considerable distances to be baptised. The prefect thus wrote to the minister of Worship, suggesting that the parish priests christen the infants in the houses they had been born in, and the minister forwarded to request to the local dioceses for an opinion.[39]

The reaction of the diocesan authorities is indicative of clergy's attitude towards state interventionism in this field. The bishop of Feltre and the vicar of Belluno pointed out that the suggestion would have created serious problems for the parish priests in their dioceses, as it would have compelled them to travel to distant villages in bad weather conditions several times a week. While acknowledging the high rates of infant mortality, the two clergymen ascribed it more to the overall winter conditions than to the practice of taking children to church to be baptised. Both the vicar and the bishop pointed out that the prefect's suggestion would have been useless unless the same measure could also be applied to the civil registration of births, since registry offices were often far away from the isolated houses these children were born in. Moreover, the vicars took advantage of the occasion to demand that the civil authorities intervene to push people to christen their children in a timely fashion during the summer months.[40] Even if, on this occasion, the prefect's proposal apparently did not overcome the objections of the diocesan authorities, a similar provision was adopted a few years later with the consent of the latter.[41] These two cases show how the clergy could react with anger to attempt by the state to intervene in its traditional registration procedures. In the second of these cases, however, the clergymen were ultimately forced to accept the government's will – even if it went against their own interests.

The interventionism of the civil authorities, however, could also offer opportunities for members of the clergy to pursue their own gain. In December 1807, the government reduced the number of parish churches in Vicenza from 16 to 10.[42] This change in the religious structure of the city ignited a conflict among the local clergy. Traditionally, every newborn in Vicenza was baptised in the city's cathedral. This centuries-old tradition provided a stable income for the cathedral archpriest, who used it to pay two other clergymen. The only time the custom had ever been interrupted had been a period of a few months during the democratic municipality of 1797.[43] However, in January 1808, new rules were adopted, openly mentioning that every new parish in the city had to have a baptistery.[44]

The parish priests took advantage of these rules to question the cathedral's pre-eminence in the field of baptism. The church of the Maddalene, a small village five kilometres from the city centre, was the first to manage to obtain a baptistery. A few months later, four other parish priests in the city asked for one. Through the prefect – who supported their request – they sent four pleas to the minister of Worship. They backed up their proposals by mentioning several negative consequences of the centralisation of christenings: it meant they had to issue certificates without being able to consult the related registers in the cathedral, while priests there, who didn't know the parents of children being baptised personally, could not be sure of the veracity of their declarations. Moreover, the trip to the cathedral could potentially harm newborns in the winter months – while sometimes their neighbourhoods were completely isolated from the city centre by floods.[45] The cathedral priests, on their part, reacted with a distressed letter to the bishop. They quoted several canonical and theological arguments to support their right to baptise all of the city's newborns, while downplaying the distance between the cathedral and the other city churches.[46] Despite this effort, the minister of Worship allowed the four parish priests to establish autonomous baptisteries in their churches.[47]

A longstanding tradition was thus overcome by a few local priests, who exploited new governmental provisions to prevail on the interests of the cathedral clergymen. To be sure, state interventionism in religious matters surely produced both cooperation and opposition among the Italian clergy. But it was also used by its members to pursue their own conflicting agendas.

Political Support and Opportunism

While they continued to manage the traditional systems of registration and identification, members of the clergy were also required to provide other sorts of active efforts for the state authorities. Every level of the Italian church hierarchy was mobilised in support of the new systems of civil identification and, most of all, registration. The clergy's responses to this call for help varied according to several factors. Starting with the Italian bishops, the clergy's behaviour should be analysed in detail to be fully understood.

As part of its policy of *ralliement* with traditional élites, the Italian government actively tried to involve the higher ranks of the clergy in its political action.[48] The *Consulta* of Lyon had proclaimed Catholicism as the state religion of the Italian Republic in 1802. The following year, the Concordat with the Holy See created the conditions for further involvement on behalf of the Italian bishops.[49] Several leading members of the clergy joined the electoral college of intellectuals and the republic's electoral body. The monarchical turn strengthened this tendency. In 1805, the new Council of State included three bishops and two other clergymen.[50] The bishops and archbishops of Milan, Ravenna, Bologna, Ferrara, and, later, Venice were – by right – Grand Officers of the kingdom, and thus members of the State Council and the Senate.[51] As of 1808, the archbishops became counts and several bishops became barons of the kingdom – the former were even given the possibility of choosing a nephew as heir to their titles.[52] Moreover, several other clergymen joined the Order of the Iron Crown – the Italian equivalent of the *Legion d'honneur*, presided over by Napoleon himself.

These appointments symbolised the will of the Napoleonic authorities to involve the higher ranks of the clergy in their state structure. However, their role was more than just symbolic. The members of the State Council, in fact, were a part of the small political élite in whose hands the kingdom's legislative power rested. The bishop of Ravenna Antonio Codronchi was the most prominent religious member of the Council. Codronchi had maintained a compromised position with the pro-French authorities since the late 1790s, and had been a pivotal actor in securing the proclamation of Catholicism as the state religion at the *Consulta* of Lyon in 1802. His pro-Napoleonic stances had prevented his promotion to cardinal by the pope, but it had led the emperor to appoint him as Grand Chaplain of the Kingdom of Italy in 1805. This new position allowed him to live in the capital until the kingdom's dissolution, and thus to easily participate in the meetings of the Council.[53]

Codronchi was present at most of the discussions on the involvement of the clergy in the civil registration system. Even if he never officially intervened in these debates, his presence probably had a role in influencing them. Two members of the clergy, Paolo Lamberto D'Allegre and Tommaso Ronna, participated more actively in Council meetings, but did not intervene when the clergy's involvement in registration procedures was being discussed.[54] In fact, the only occasion in which the religious members of the Council intervened on this regard was quite a special one. In May 1805, the viceroy Eugene was in Milan to oversee the transformation of the Italian republic into a monarchy. Ten days before Napoleon's coronation as king of Italy, he took advantage of a discussion in the Council to intervene with a proposal to delegate the management of the civil registers of births and deaths to parish priests. On this occasion, both Codronchi and the archbishop of Milan, Giovanni Battista Caprara, spoke in favour of the viceroy's proposal, pointing out, nevertheless, that parish priests could never be required to civilly register marriages that were not allowed under canon law.[55] Even if their presence was discreet, the bishops were at the very heart of the Italian institutions – where they could make themselves heard by the Napoleonic élites.

Still, Italian bishops did have to give something in return for the special status they enjoyed in Napoleonic Italy. Their privileges depended on their unwavering support for the government, but they were also required to intervene directly in backing some of the more contentious policies promoted by the Milanese authorities. Since 1804, for instance, the Italian clergy was heavily involved in promoting smallpox vaccination: the bishops coordinated and encouraged this involvement across the national territory.[56]

A similar kind of cooperation is also visible in the circulars that bishops sent their parish priests on the subject of civil registration. Like the circular analysed at the beginning of this chapter, these documents mainly originated from requests by the Italian government. The minister of Worship, Bovara – himself member of the church – sent at least eight circulars on civil registration to all the kingdom's bishops between 1806 and 1811.[57] On other occasions, Bovara targeted specific territories, asking the bishops there to send circulars to their priests.[58] The ultimate aim of these many circulars was to get the parish priests to follow the rules and explain them to the faithful. Every one of Bovara's circulars was followed by specific circulars in every one of the kingdom's dioceses, in which the local bishops reinterpreted and tried to explain the government's stance. In these texts, the bishops often went above and beyond their duty of reminding the priests of the rules of civil registration, often providing theological and political arguments in favour of the government's policies.

The circular by the bishop of Udine analysed at the beginning of this chapter was thus only one of the dozens of texts sent by Italian bishops to their parish priests to explain the importance of civil registration in early 1809. The length of the circulars went from the one-page summary penned by the bishop of Lodi – Gianantonio Della Beretta – to the elaborate four-page composition sent by the bishop of Crema, and member of the State Council, Ronna.[59] The bishops' circulars generally adopted a paternalistic tone towards the general population. In these circulars, priests were depicted as the shepherds of their communities, and were thus encouraged to enlighten the "ignorant" populace. According to the bishop of Modena Tiburzio Cortese, the population 'normally sees and considers only the present day, while it very easily neglects or completely disregards everything that can happen in the future'.[60] This paternalistic attitude was also displayed in the bishops' relationship with the priests themselves. The bishop of Bergamo Giovanni Paolo Dolfin, for instance, called his priests 'docile lambs', who would easily listen to their shepherd (the bishop himself). It was their duty to 'destroy the ignorance of the people entrusted to your spiritual care'.[61]

A large number of these circulars was devoted to making sure the priests – and indirectly to the population at large – knew the reasons why civil registration duties had to be respected. First of all, the bishops mentioned religious and even theological reasons. As the bishop of Vigevano Francesco Milesi explained, 'a Catholic [has the duty] to obey sovereign laws promptly and exactly. [...] The submission to sovereigns in everything that is under their jurisdiction is a part of god's law and of Christian piety'.[62] The need to submit to the civil authorities was also sustained by references to the Bible. Milesi referred to the famous gospel passage 'render unto Caesar the things that are Caesar's', while the more sophisticated archbishop of Udine mentioned 'the great doctrine of the apostle Peter (I Peter [2:]11–13), that is, that for the love of god we are to be subject to any man who has authority over us'. The bishop of Vigevano summarised these arguments by writing that 'obeying god implies a total, filial, constant, unchangeable submission to the sovereign'. Nevertheless, references to the obedience to political power remained, in most cases, general, and did not mention the current government. Only the bishop of Cittanova, Teodoro Laureatano Balbi, went so far as to openly celebrate the 'provident and healthy laws of our august and very wise sovereign Napoleon the Great, who is the vice-regent on earth of the superior celestial dominion, tasked with providing public and private happiness'.[63] Most of the bishops, however, simply stressed that Christians had to obey political power, whichever way this was construed. One of these was the archbishop of Brescia, Gabrio Maria Nava, who pointed out that 'every power comes from god, and anyone who resists it, resists god himself'.[64]

If the submission of organised religion to political authority was necessary in all respects, it was even more important in the case of civil registration. The bishop of Como, Carlo Rovelli, mentioned the 'angelic doctor saint Thomas [Aquinas]' in his circular, arguing that the medieval philosopher had been the first to recognise civil jurisdiction on marriage.[65] The archbishop of Udine went even further. In his circular, he pointed out that if Quirinus had not imposed a census on Judea, Jesus would have not been born in Nazareth as the prophets had announced. The biblical story of the birth of Jesus Christ was, in this way, mobilised to provide a (religious) justification for the importance of civil registration.

Still, in these circulars, religious arguments in favour of the new legislation took up less space than other, more practical justifications. Most of the bishops felt the need to clarify the long-term importance of the civil effects of registration. The bishop of Reggio, Francesco Maria d'Este clearly stated that the civil registers were going to become the only legal sources for settling problems related to contracts and inheritances, 'for which disputes and obstinate fights often arise, at great expense and causing very confused discussions'.[66] These explanations were repeated several times by the bishops, who highlighted the possible negative consequences of foregoing registration for marriages, inheritance procedures, and children. The bishop of Verona, Innocenzo Maria Liruti emphasised that the civil registers were going to be 'necessary to preserve the public and private rights and, regarding marriages, the basis of the civil legitimacy of wedding vows and their offspring'.[67] According to the bishop of Modena, 'every right of every man and every family in society' would depend on civil registration, since the 'the ambiguous and dangerous evidence of mere witnesses' was not sufficient to guarantee civil acts. He clearly spelled out the future consequences of not taking civil registration seriously.

> The future ages will one day protest against the crimes of their ancestors; and maybe also an individual man, finding himself in a changed state of fortune or interest or condition, might suffer some harm from the damage caused to him by his omission from the registers. And who is it that can say of himself and of his children, for his property, 'I will never have need to resort to the registers?'.

The Italian bishops had a crucial role in elaborating and spreading political discourses in support of civil registration. In their circulars, they explained the religious, theological, and practical reasons for which parish priests had to cooperate with government policy. The efforts that individual bishops put into this, however, varied wildly. As mentioned, one of the longest and most heartfelt circulars was written by Tommaso Ronna, who had benefited greatly from the Napoleonic regime. He had started his career as the parish priest of the Milanese basilica of San Babila, before being made a member of the Council of State and promoted to the bishopric of Crema. One of the shortest and most unenthusiastic circulars, on the other hand, was penned by Giovanni Antonio della Beretta, the bishop of Lodi who had been a staunch opponent of the pro-French governments since the 1790s.[68] The bishops' level of engagement in support of the new policies naturally depended on their personal attachment to the government. However, they all contributed – even the less enthusiastic ones – to explaining the new procedures of civil registration and building a consensus around them.

Besides their words, the practical actions of the Italian clergy confirmed that different levels of support existed among them. In August 1806, for instance, the old vicar of Brescia Pietro Angelo Stefani – who was acting as temporary bishop until the appointment of a new one – issued a circular to address the issue of religious marriages that had been celebrated without previous civil registration – as prescribed by law. In the words of the vicar, this was a 'shameful disorder' that would stop 'civil and political laws and religion from going hand in hand and walking fraternally, conspiring jointly for the common happiness and prosperity'. The vicar therefore ordered that no marriage could be celebrated without a certificate from the episcopal curia – which, in turn, could only be issued upon presentation of a civil certificate of marriage. Moreover, he tasked the curia with compiling a list of all the civil marriage certificates used to issue these new certificates.[69]

In doing so, the vicar was using his pastoral power not only to enforce civil registration, but also to exert a stricter level of control over the actions of his parish priests. This excessive zeal was not appreciated by the local prefect, who wrote to the minister of Worship and the minister of Justice to complain. The prefect did not approve of the vicar's circular for logistical reasons. He believed the new requirement would create travel and stamp costs that would delay the celebration of religious marriages. The prefect had also received complaints from some parish priests, and especially from those living in villages and towns far from Brescia.[70] For once, the roles were reversed. The vicar had wanted to create a new system that would have virtually eliminated the discrepancies between civil and religious marriages, but the prefect opposed these rules in the interest of making life easier for citizens. At the same time, he also acted as a conduit for the irritation of some parish priests.

Nevertheless, the bishops could also use their agency to serve religious purposes. If parish priests could be urged to remind their parishioners to celebrate civil as well as religious marriages, civil registrars could also be encouraged to make sure that the couples they married would have a religious ceremony, too. This was a strategy adopted, for instance, by the bishop of Concordia, Giuseppe Bressa, in June 1811. After issuing a circular to his local priests in which he asked them to urge couples to celebrate a civil ceremony only after they had everything in order for a religious one, the bishop wrote to the prefect of the Adriatico with a counterproposal. It was not uncommon for couples to celebrate civil marriages even if canonical impediments did not allow them to have a church marriage. Therefore, some civil marriages were not 'confirmed' with a subsequent religious ceremony. The bishop therefore proposed that the registrars ask prospective spouses whether everything was in order for their religious ceremony before celebrating their civil marriage. He also mentioned the possibility of establishing specific parish certificates which would allow couples to proceed with civil marriage.[71] The bishop's proposal reversed the usual terms of cooperation between civil and religious authorities. Generally, the civil authorities took advantage of religious authorities to help enforce civil registration. In this case, instead, a religious authority asked for help in enforcing canon law.

As the case of the Udine bishop – discussed at the beginning of this chapter – demonstrates, the support from leading members of the clergy towards civil registration could be quite ephemeral. During the brief 1809 Habsburg invasion, Giuseppe Bressa, the same bishop of Concordia mentioned above – who had previously given no significant signs of opposition to the Napoleonic authorities – slightly but substantially changed the text of the document allowing couples to proceed with religious weddings in his diocese. While the previous wording explicitly mentioned the duty to hold a civil ceremony before the religious one, this sentence promptly disappeared as the Napoleonic authorities evacuated the occupied territories.[72] More than a political statement, however, this subtle change was probably part of the bishop's rapid adjustment to the change of regime. The bishop of Concordia was simply adapting his role to the new authorities in town, as he had previously done with the establishment of the democratic municipality in 1796, the creation of the first Habsburg administration in 1797, and the return of the pro-French government in 1805.[73]

In this sense, Bressa exemplifies the stances of several Italian bishops. Like many of his colleagues, he maintained a pragmatic attitude toward the pro-French governments and their provisions on civil registration. He and several of his colleagues remained in a grey zone for most of the Napoleonic period, eschewing both active support and implicit opposition. They did sustain the government's policies of state building, even if unenthusiastically. But their support was also frail, and crumbled at the first hurdle.[74] Nevertheless, regardless of their inner convictions, the bishops' role in politically and practically supporting the establishment of the civil registration system cannot be underestimated.

Registering Priests

At the beginning of 1811, the municipality of Grado tried to appoint the parish priest of Belvedere as the adjunct registrar of his village. Belvedere was across the Marano lagoon from the municipal seat, making connections with Grado difficult. Despite the manifest necessity of bringing the civil registers closer to the local population, the priest refused this appointment. As he wrote to the local vice-prefect, he was 'lonely in the parish' and he often had to leave the village to cure diseases caused by the 'unhealthy' air. He thus proposed that a municipal official take on the position, especially since the village did not enjoy any of the services provided in the municipality's main centre. It would somehow have been an injustice, he argued, to appoint him as adjunct registrar.[75]

The priest managed to avoid being appointed, but he established some form of cooperation with the municipal registry. Every time a religious ceremony took place in Belvedere, he communicated the details of it to the registrar in Grado, who used them to inscribe the relative acts in the civil register.[76] This highly irregular practice was not the only form of involvement that the priest of Belvedere had in civil registration procedures. In July 1812, he reported several mistakes in the conscription lists drawn up for the village. 'Pushed only by [his] zeal for public service and the conscripts' good', the priest detailed 27 mistakes (including the fact that ten local young men were absent from the lists).[77] A few months later, the vice-prefect once again asked the parish priest take up the role of adjunct registrar in Belvedere. Once again, he refused.[78]

Just like the bishops, Italian priests also displayed a wide variety of attitudes and behaviours with regard to the new civil registration procedures. As the case of Belvedere shows, the same clergyman could refuse the official role of adjunct registrar for his village while still cooperating with the authorities on a more informal level. Indifference, however, was never an option. Italian priests not only continued to perform the religious identification and registration practices of the past; they were also actively involved by the Italian government in several official activities. Their position was officially inscribed in the state's bureaucracy. In fact, it was the government that paid them, keeping an eye on them via local cantonal delegates of the ministry of Worship. The government also constantly asked them to cooperate in supporting its policies. Their reactions, however, varied considerably.

In August 1806, the parish priest of the small village of Valgreghentino, Fortunato Brini, wrote a letter to the interior minister. In the first few months since the implementation of civil registration, he said, he had noticed that several births and deaths had not been registered in the civil registers of neighbouring villages. He thus proposed to change the regulations and make the civil registration mandatory in order to perform religious christenings and burials – following the rules established for marriages. In his letter, he highlighted that he was moved by his 'zeal for the perfect compliance with the sovereign orders' and by a will to avoid any future negative consequences for those who neglected their civil registration duties.[79]

Brini's pro-government stance did not change even after the dramatic rupture between the Napoleonic authorities and the Church in 1809. In October 1811, he took advantage of a national survey of local traditions to propose the composition of 'beautiful patriotic songs to be spread among the population to promote the respect to the supreme being and the loyalty to the king and the fatherland – which are the same thing'.[80] Father Brini was not the only Italian priest to show an attachment to the government that went beyond mere submission to political power. In June 1809, for instance, the parish priest of Meolo responded enthusiastically when the authorities asked him to explain the importance of civil registration to his parishioners during mass. As the local mayor reported, the priest had urged the faithful, from the altar, to pay 'blind obedience' to the government's provisions.[81]

Parish priests were especially important for the establishment and management of civil registration in rural and peripheral areas. This was the case, for instance, in the territories annexed to the department of the Adige in 1806. According to an estimate made by the local prefecture, parish priests would be appointed as civil registrars in 38 of the 90 new municipalities that were going to be set up.[82] The importance of parish priests for the organisation of these new municipalities is confirmed by the list of civil registrars operating in the department of the Adige between 1806 and 1807. A total of 88 civil registrars were appointed in this period in the municipalities annexed in 1806. At least 18 of these – one-fifth of the total – were parish priests.[83]

The clergy had the same role in supporting the civil registration system in the kingdom's mountainous departments. In the department of the Piave, for example, the local authorities struggled for years to establish an efficient civil registration system.[84] As the local prefect explained in December 1806, most of the municipal officials in the Cadore, an alpine district, emigrated each winter to other Venetian departments or to Habsburg territory, leaving only women, children, and old men in the villages. Those who stayed behind could not, according to the prefect, be appointed as registrars. They were unable to read and – instead of writing – they were used to noting down symbols with pieces of coal on planks of wood or on walls.[85] In 1810, the prefect opted for a drastic solution to solve the problem: he decided to ask the local parish priests to take care of civil registration in the district. To his surprise, they all accepted his proposal. Since they had already been ordered to christen babies in the houses they had been born in during the winter months, carrying out civil registration tasks did not seem like that much of an extra burden.[86] From the point of view of the parish priests in the Cadore, civil registration duties were only a small addition to their traditional tasks.

Nonetheless, parish priests also played an important role in departments in the lowlands, like the department of the Adriatico. Here, as in other places across the kingdom's territory, parish priests were often the only possible choice when it came to appointing adjunct registrars in rural villages. In December 1808, for instance, the municipality of Bottrighe proposed a parish priest and a chaplain as adjunct registrars because they were the only available individuals in their villages.[87] Similarly, in 1811, the mayor of Aquileia proposed that the adjunct registrar of Scodovacca, who had resigned, be replaced by the local parish priests, who was 'the only [person] capable' of carrying out civil registration tasks.[88] The lack of alternatives remained a structural issue right into the final years of Napoleonic rule. In January 1812, for example, the registrar of Portogruaro proposed two parish chaplains as adjunct registrars in their villages, explaining that he had no one else to put forward for the job because there were no other possible candidates.[89]

Nevertheless, this reliance on parish priests did not just stem from a lack of viable alternatives. The mayor of Fossalta, for instance, thought that nobody would have made the trip to the municipal seat to have births inscribed in the civil register if the parish priests in the surrounding villages were removed from their position as adjunct registrars.[90] Even when not indispensable, parish priests were still convenient registrars because they usually received no payment for taking on registration tasks. When, in August 1812, the mayor of San Stin was asked to propose a new adjunct registrar for the village of Torre di Mosto who could replace the local parish priest, he found himself in a difficult position. As the mayor explained to the prefect, there were other candidates in the village, but none of them would have accepted the job without a salary. The problem was that the municipal budget could not support such an expense. The parish priest, for his part, was receiving no payment for his role as adjunct registrar.[91]

Even in those localities in which they were not in charge of the system, priests still had a role in supporting civil registration. In the first year after its implementation, the registration system of the municipality of Torcello was in serious financial and logistical trouble. The municipal territory was too big, and transport across it was impeded by the Venetian lagoon. The registrar had therefore appointed seven unpaid assistant registrars in order to make registration procedures more convenient for the local population. However, several assistants turned out to be incapable of managing the civil registers without any help. The registrar in Torcello thus had to ask the parish priests of the seven villages in which the assistants operated for active assistance.[92] However, a new problem then arose, which also led the municipality to enlist the help of the clergymen. Even if the assistants were unpaid, the municipality could not afford to buy the stationery needed to establish the new local registers. Therefore, the registrar in Torcello asked the local priests to persuade their parishioners to contribute 20 *soldi* per household to cover these expenses. He also pointed out how convenient it would be for the villagers to have registries closer to home. Despite the priests' intervention, however, 'very few' villagers contributed.[93]

The involvement of parish priests, however, was no guarantee that the civil registers would be kept in good order. In April 1811, for instance, the local justice of the peace discovered that the civil registers of two villages in the municipality of Caorle were being poorly managed by the priests who had been appointed as adjunct registrars there. Informed by the justice of peace, the mayor of Caorle wrote to the vice-prefect to prevent him from reprimanding them. He pointed out that the registers had, in previous years, been managed by other inhabitants of the villages, with similarly disappointing results. The civil registers had therefore been entrusted to the priests.[94] According to the mayor of Caorle, the priest of the village of Cottoni compiled the registers with a completely wrong layout and accepted witnesses who were under the legal age. The priest of San Giorgio, meanwhile, was generally 'unable' to manage the registers. In both cases, however, there were no available alternatives.[95] The vice-prefect then wrote directly to the priests urging them to stop this 'reprehensible irregularity'. He stated that he was no longer willing to tolerate any further 'disorder' and that he was ready to use any means in his power to prevent it.[96]

The relationship between local authorities and parish priests was further complicated by the fact that cooperation from the latter could not be taken for granted. In late 1806, for instance, several priests in the department of the Olona refused to be appointed as civil registrars in the small villages they resided and worked in.[97] Even when they did accept to act as adjunct registrars or registrars, members of the clergy could interrupt this cooperation at any time. This was the case in early 1807, when several parish priests in the department of the Adda took advantage of a circular sent by the minister of Worship Bovara to refuse any further involvement in the civil registration system.[98]

However, most these resignations were usually individual. In cases such as these, the civil authorities usually tried to change the mind of the individual priests who resigned. This is what happened to a chaplain in Annone who, in 1811, had resigned from the position of adjunct registrar less than one year after taking on the job. The mayor tried to convince him to stay on, but the chaplain remained 'adamant'.[99] Sometimes, municipal authorities tried to make this cooperation more advantageous for the clergymen. In December 1812, the mayor of Cinto asked for authorisation to exempt two local priests who acted as adjunct registrars from paying municipal taxes, as a reward for their service. He hoped, in this way, to push them to withdraw their resignations – especially since he would have been unable to find anyone else to act as adjunct registrar in these villages.[100]

There were various reasons why priests could resign. The priest of Teglio, in the municipality of Cinto, mentioned his 'inaccuracy' in the management of the civil registers as the cause for his resignation. In the previous two years, he had remained silent – he wrote in his plea to the vice-prefect – because of his 'deference to authority'. He had only decided to break his silence because he had learned that the prefect was trying to replace priests in the department who acted as adjunct registrars and, more importantly, because he had found a willing substitute for the position.[101]

Similarly, the parish priest of Torre di Mosto wrote directly to the prefect of the Adriatico to ask to be removed from the position of adjunct registrar which he had accepted one year and three months earlier. In his plea, he wrote that he had had to use an assistant – compensated with 'some tip from time to time' – because he was 'completely unaware of the required practices, and old, and idle, and his nervous problems made him completely unable to [hold] this office'. In the previous months, he had lost his assistant and the pain in his joints had worsened. Moreover, the priest added that the municipal registrar was 'handsomely paid' – and suggested that the registrar's father would be a perfect substitute as adjunct registrar for the village. The priest also stressed that he was not looking for financial retribution. He thought that the prefect would not have allowed for there to be an unpaid adjunct registrar in the village when there was already someone (namely the registrar) who 'without effort and disturbance accepts a salary'.[102] In this case, the priest's resentment of the local municipal authorities played just as important a role in his decision to resign as the physical difficulties caused by his old age.

The many different ways in which parish priests behaved during the brief Habsburg occupation of 1809 offer a confirmation of the variety of attitudes that the lower ranks of the clergy displayed towards civil registration.[103] During the occupation, parish priests in several municipalities performed marriages and burials without checking if these had previously been noted in the civil registers. In the district of Feltre, for example, between 20 April and 20 May, parish priests celebrated one marriage and performed seven burials without requiring proof of civil registration.[104] Some clergymen, however, also adopted more prudent attitudes during this exceptional period. In the hilly village of Salcedo, for instance, the parish priest continued to ask for civil registration certificates before celebrating marriages and carrying out burials, even if the mayor and civil registrar saw the practice as obsolete now that they were no longer under Napoleonic rule.[105]

As these last cases confirm, parish priests held mixed and ambiguous attitudes to civil registration. In several areas of the Kingdom, they were essential in setting up and managing the civil registers as adjunct registrars and registrars. Their cooperation, however, was highly situational and often ephemeral. They cooperated with civil authorities only until their interests were not harmed by the required effort. Even if the Napoleonic authorities often needed their support, this was not always guaranteed.

*

The Italian clergy was mobilised at every level to support the new identification culture that the government was promoting. At the same time, the civil authorities slowly broadened their jurisdiction to cover some of the traditional procedures of identification and registration managed by members of the church. This twofold action was consistent with the Napoleonic understanding of the clergy as part of the state machine. However, even if bishops and other leading clergymen were co-opted as part of the kingdom's political élites, the reactions to state interventionism at different levels of the church hierarchy varied widely.

In a general sense, the cooperation of both the bishops and the parish priests was crucial for the establishment of the new state system of identification. Without their political and practical support, it would not have been possible to implant the civil registration system in large parts of the Kingdom of Italy's territory. The cooperation between church and state was surely inspired by the traditional Catholic doctrine of submission to political power. Especially among the bishops, this collaborative attitude was also accompanied by a certain degree of political opportunism. Overall, the Italian clergy proved to be a valuable ally in the construction of new administrative systems inspired by the state culture of identification.

However, if we take a closer look, this cooperation reveals more complex traits. A variety of different behaviours existed beneath the clergy's nominal support for the identification system. The only attitude that the state's interventionism did not allow was indifference: at every level, the clergy was compelled to either cooperate with or resist the new procedures. Even if open acts of resistance or opposition were rare, clergymen showed that their cooperation was far from being unwavering or wholehearted. The political support of bishops often vanished whenever the Napoleonic authorities evacuated territories occupied by the Habsburg Empire. Even in areas that stayed under Napoleonic control, the fleeting practical cooperation of parish priests had to be constantly renegotiated with the civil authorities.

Members of the clergy, however, did maintain one red line in their dealings with the government: the respect of the canon law. The new civil procedures were not met with widespread opposition from the Italian clergy because they did not contradict its own traditional culture of identification. For centuries, clergymen had managed an identification and registration system that had similar features to the one promoted by the Napoleonic state at the beginning of the nineteenth century. They were both based on the idea that each individual had a single identity, and that this "real" identity had to be certified by written acts. It was no coincidence that religious authorities often asked their civil counterparts to support their own procedures of identification too: this was an implicit acknowledgement that the two systems were based on similar principles. However, the coexistence of these two systems became problematic when the Napoleonic authorities tried to intervene in fundamental aspects of the clergy's own culture of identification. The friction between canon law and civil legislation was just the consequence of a deeper divergence opening up between two different cultures of identification. This is why the Italian clergy always remained an ambiguous partner for the Napoleonic state in the field of personal identification: its much-needed support was always conditioned on its need to respect its own culture of identification.

Notes

1 Circular from the minister of Worship to the bishops (31 December 1808, n. 14206), ASMi, Popolazione p.m., b. 5.
2 Circular from the archbishop of Udine to the parish priests (20 January 1809), *ibid.*
3 Giovanni Miccoli, "La Chiesa friulana fra occupazione francese, dominazione austriaca e regno italico," in *Opere d'arte di Venezia in Friuli*, ed. Gilberto Ganzer (Udine: Magnus, 1987), 58.
4 Anna Maria Rao, "Napoleonic Italy: Old and New Trends in Historiography," in *Napoleon's Empire: European Politics in Global Perspective*, ed. Ute Planert (New York: Palgrave Macmillan, 2015), 84–97. Notably, Michael Broers highlighted the conflictual relationship between the two institutions, focusing in particular on the departments annexed to the French Empire: Michael Broers, *The Politics of Religion in Napoleonic Italy. The War Against God, 1801–1814* (London-New York: Routledge, 2002), in particularly 175–189.
5 On religious reforms in the Kingdom of Italy: Carmelo Amedeo Naselli, *La soppressione napoleonica delle corporazioni religiose. Contributo alla storia religiosa del primo Ottocento italiano* (Roma: Editrice pontificia università gregoriana, 1986); Filiberto Agostini, *La riforma napoleonica della chiesa nella repubblica e nel regno d'Italia 1802–1814* (Vicenza: Istituto per le ricerche di storia sociale e religiosa, 1990); Daniele Menozzi, "L'organizzazione della chiesa italiana in età napoleonica," *Cristianesimo nella storia* 14 (1993): 69–97. On the incarceration of Pius VII: Ambrogio A. Caiani, *To Kidnap a Pope: Napoleon and Pius VII* (New Haven, CT-London: Yale University Press, 2021), 1–9 and 132–249.
6 Claudio Canonici, "'Per non abbandonare la Chiesa né il popolo'. Il giuramento ecclesiastico negli "Stati romani" in epoca napoleonica (1810–1814)," *Rivista di storia del cristianesimo* 1/2 (2004): 303–331; Claudio Canonici, "Clero e giuramenti," in *Italia napoleonica. Dizionario critico*, ed. Luigi Mascilli Migliorini (Torino: Utet, 2011), 107–109.
7 Antonia Pasi, *Contare gli uomini. Fonti metodi temi di storia demografica* (Milano: LED, 1992), 49–53.
8 See for instance: copy of a circular from the bishop of Vicenza to the parish priests of the city and the vicars outside the walls (28/104/1787), Diocesan Archives of Vicenza, Actorum, b. 252.
9 For the Duchy of Milan, see for instance: circular of the *magistrate camerale* (3 March 1779), ASMi, Popolazione p.a., b. 16.
10 Report from the Interior minister to the Executive Directory (14 December 1797, 24 frimaio VII), *ibid.*
11 'Regolamento per la Polizia generale' (27 March 1802), in *Bollettino*, 1802, 41–47.
12 Roberto Bizzocchi, "Marchigiani senza cognome. Un'inchiesta nell'Italia napoleonica," *Quaderni storici* 134 (2010): 547.

13 Copy of the instructions for compiling the general list of the population in Venice (undated, n. 9678), ASMi, Popolazione p.m., b. 70.
14 Announcement by the archbishop of Milan (24 September 1801, 2 vendemmiale IX), in *Raccolta*, 2/III, 73.
15 Draft letter from the vice-prefect of Chioggia to the bishop of Chioggia (15 June 1808, n. 1683), ASVe, Chioggia, b. 39.
16 Letter from the prefect of the Adriatico to the vice-prefect of Chioggia (1 June 1808, n. 10812). The vice-prefect ultimately granted the bishop the authorisation to correct the year of birth and alerted the municipality: copy of a letter from the vice-prefect of Chioggia to the bishop of Chioggia (23 July 1808, n. 2341); copy of a letter from the vice-prefect of Chioggia to the municipality of Pellestrina (23 July 1808, n. 2341), all in *ibid.*
17 'Legge generale di Finanza per gli anni 1805 e 1806' (17 July 1805), in *Bollettino*, 1805, 425–447.
18 See, for instance, Tommaso Negri's parish certificate of good behaviour: 'I, the undersigned, certify and attest that Tommaso Negri son of Giuseppe is a parishioner of good conduct, and thus I believe him to be incapable of disrupting the public peace' with a short note from the municipality: 'Seen by the Vicenza municipality': ASVi, Commissariato, b. 6, f. 540.
19 Circular from the minister of Worship to the prefects (9 June 1807, n. I/6386), ASBo, Prefettura, 1807, tit. XX, rub. 10.
20 Sec. XII, art. 152, subsec. III, art. 8 of the 'Decreto sulla tariffa dei diritti delle ipoteche, sulla carta bollata e sul registro' (21 May 1811), in *Bollettino*, 1811a, 325–416.
21 Letter from the justice minister to the interior minister (9 November 1811, n. II/19707), ASMi, Popolazione p.m., b. 7. The following year, the government declared that municipal endorsements on religious certificates were also exempt from stamp duty: draft circular from the interior minister to the prefects (22 June 1812, n. II/16717), ASMi, Popolazione p.m., b. 8.
22 Plea from Paolo Mazzucchelli to the Police Court of Milan (undated, but October 1798), ASMi, Culto p.a., b. 2222.
23 For instance: plea by Maria Teresa Frascinedi to the prefect of the Reno (3 December 1807), Prefettura, 1807, tit. XX, rub. 11.
24 'Regolamento per l'attivazione in tutto lo Stato dei Registri delle nascite, de' matrimonj e delle morti' (27 March 1806), in *Bollettino*, 1806a, 168–223.
25 All women and men over 60 years of age were exempt from this duty: announcement by the prefect of police of the Olona (20 March 1811), ASMi, Popolazione p.m., b. 7.
26 Letter from the prefect of the Agogna to the minister of Worship (20 March 1811), ASMi, Culto p.m., b. 86. The use of pre-printed forms was not a novelty in early nineteenth-century Italy. Already in the 1760s, the diocese of Milan used pre-printed forms for authorisation requests for parish certificates: see different authorisation requests (1765) in ASMi, Culto p.a., b. 2222.
27 On similar trials in an earlier period: Teresa Bernardi, "Control of Marriage and (Im)mobility in Venice (16th–17th Centuries) in Dialogue with Cultures of Vigilance," *Società e Storia* 177 (2022): 546–568.
28 Plea by Anna Maria Dalla Presepia (undated, but 07/1810); verdict of the general vicar of Vicenza (14 July 1810), both in Diocesan Archives of Vicenza, Processi battesimi e morti 1791–1816, b. 816.
29 Plea by Antonio Martini and Pasqua Felice Rosa (undated, but 02/1811); interrogations by the vicar-general of Vicenza (2/1811), both in *ibid.*
30 Copy of a letter from the municipality of Lodi to its secretary (16 August 1803, n. 1870); draft letter from the minister of Worship to the municipality of Lodi (13 September 1803, n. 9987), both in ASMi, Culto p.m., b. 86.

31 Letter from the vicar-general of Verona to the minister of Worship (23 June 1806), ASMi, Popolazione p.m., b. 52.
32 Confidential letter from the prosecutor of the Olona to the justice minister (6 October 1806, n. 1935); draft letter from the justice minister to the prosecutor of the Olona (7 October 1806, n. 296/10188), both in *ibid.*
33 Concordat between the Italian Republic and the Holy See (26 January 1804), in *Bollettino*, 1804a, 42–55.
34 Confidential letter from the prosecutor of the Tagliamento to the justice minister (28 August 1806, n. 621), ASMi, Popolazione p.m., b. 52.
35 Letter from the registrar of Chioggia to the prefect of the Adriatico (13 February 1807, n. 26), ASVe, Prefettura, b. 90.
36 The vicar also argued that, in his opinion, the priest had not wanted to break the law and that he would not have made these mistakes if had he been duly informed: letter from the vicar capitular of Chioggia to the prefect of the Adriatico (9 March 1807), ASVe, Prefettura, b. 90.
37 Circular from the president of the Central Health Authority to the parish priests of the department of the Olona (27 January 1806). The same circular was later sent to all the bishops of the kingdom: draft circular from the minister of Worship to the bishops (1 February 1806, n. 965), both in ASMi, Culto p.m., b. 86.
38 Letters from the parish priest of the church of San Babila (undated), the parish priest of the church of Santo Stefano (28 January 1806), the parish priest of the church of San Nazario (29 January 1806) to the central health authority, all in *ibid.*
39 Copy of a letter from the prefect of the Piave to the interior minister (9 February 1807), *ibid.*
40 Letter from the bishop of Feltre to the minister of Worship (12 March 1807); letter from the vicar of Belluno to the minister of Worship (18 March 1807), both in *ibid.*
41 Letter from the prefect of the Piave to the interior minister (21 February 1810, n. 1160), ASMi, Popolazione p.m., b. 80.
42 Decree (18 December 1807), in *Bollettino*, 1807c, 1211–1213.
43 Copy of a letter from the president of the municipality of Vicenza to the same municipality (2 July 1797, 14 messidoro 5), ASMi, Culto p.m., b. 86.
44 Copy of a letter from the minister of Worship to the bishop of Vicenza (13 January 1808), *ibid.*
45 Pleas from the parish priests of the churches of San Marco, Santa Croce, San Pietro, and San Silvestro to the minister of Worship (undated, but 10/1808); letter from the prefect of the Bacchiglione to the minister of Worship (23 October 1808, n. 19435), all in *ibid.*
46 Letter from the clergymen of the Vicenza cathedral to the bishop of Vicenza (undated, but October 1808), forwarded by the bishop of Vicenza to the minister of Worship (17 October 1808), both in *ibid.*
47 Draft letter from the minister of Worship to the prefect of the Bacchiglione (8 November 1808, n. 11675), *ibid.*
48 On the policies of *ralliament* and *amalgame* – and on their limits – in Napoleonic Italy, see Broers, *The Napoleonic Empire in Italy*, 123–212. On bishops in the Kingdom of Italy: Agostini, *La riforma napoleonica della chiesa nella repubblica e nel regno d'Italia*, 180–205.
49 On the positive attitude of Italian bishops after the 1803 Concordat, see Daniele Menozzi, "La Chiesa, la rivoluzione francese e l'impero napoleonico," in *Storia della società italiana. L'Italia giacobina e napoleonica*, ed. Daniele Menozzi (Milano: Nicola Teti, 1985), 173–174.
50 Decree (9 May 1805), in *Bollettino*, 1805, 79–80.
51 Constitutional statute (29 March 1805), in *ibid.*, 54–63.

52 Constitutional statute (21 September 1808), in *Bollettino*, 1808b, 824–847.
53 Giuseppe Pignatelli, "Codronchi, Antonio," in *Dizionario Biografico degli Italiani*, vol. 26 (1982), 599–604.
54 Both D'Allegre and Ronna were also members of the Council of Auditors: decree (9 June 1805), in *Bollettino*, 1805, 305–306. In 1806, they were appointed respectively bishop of Pavia and bishop of Crema: decree (15 April 1806), in *Bollettino*, 1806a, 308; decree (19 July 1806), in *Bollettino*, 1806b, 773.
55 Session of the State Council (13 May 1805), ASMi, Popolazione p.m., b. 5.
56 Grab, "Smallpox Vaccination in Napoleonic Italy," 38–58.
57 Circular (15 March 1806); circular (26 July 1806), mentioned in the circular from the diocesan vicar of Brescia (11 August 1806), both in ASMi, Popolazione p.m., b. 52; circular (28 October 1806); circular (22 December 1806), both in in ASMi, Popolazione p.m., b. 4; circular (31 December 1808), ASMi, Popolazione p.m., b. 5; circular (8 October 1810), ASMi, Popolazione p.m., b. 54; circular (22 December 1810), ASMi, Popolazione p.m., b. 6; circular (29 March 1811), ASMi, Popolazione p.m., b. 7.
58 In 1808, for instance, Bovara asked the diocesan authorities of Belluno and Udine to send a circular to their parish priests on the mandatory civil registration of newborns: letter from the diocesan vicar of Belluno to the minister of Worship (27 March 1808); letter from the archbishop of Udine to the minister of Worship (25 May 1808), both in ASMi, Popolazione p.m., b. 4.
59 Circular from the bishop of Crema to the parish priests (7 February 1809); circular from the bishop of Lodi to the parish priests (27 February 1809), both in ASMi, Popolazione p.m., b. 5.
60 Circular from the bishop of Modena to the parish priests (14 January 1809, n. 82/83), *ibid.*
61 Circular from the bishop of Bergamo to the parish priests (20 January 1809), *ibid.*
62 Circular from the bishop of Vigevano to the parish priests (11 January 1809), *ibid.*
63 Circular from the bishop of Cittanova to the parish priests (17 January 1809), *ibid.*
64 Circular from the bishop of Brescia to the parish priests (19 January 1809, n. 78), *ibid.*
65 Circular from the bishop of Como to the parish priests (10 January 1809), *ibid.*
66 Circular from the bishop of Reggio to the parish priests (7 February 1809), *ibid.*
67 Circular from the bishop of Verona to the parish priests (10 January 1809), *ibid.*
68 Della Beretta was nevertheless appointed baron by the government: Costanza Ichino Rossi, "Della Beretta, Giovanni Antonio," in *Dizionario Biografico degli Italiani*, vol. 36 (1988): 691–694.
69 Circular from the episcopal vicar of Brescia to the parish priests (11 August 1806), ASMi, Popolazione p.m., b. 52.
70 Confidential letter from the prefect of the Mella to the justice minister (26 August 1806, n. 554); confidential letter from the prefect of the Mella to the minister of Worship (26 August 1806, n. 554), both in ASMi, Popolazione p.m., b. 52.
71 Letter from the bishop of Concordia to the prefect of the Adriatico (23 June 1811, n. 182), ASVe, Prefettura, b. 383.
72 Copy of order from the bishop of Concordia (4/1808); copy of order from the bishop of Concordia (21 April 1809), both in ASMi, Popolazione p.m., b. 70.
73 On Bressa: Pietro Zovatto, "Bressa, Giuseppe Maria," in *Dizionario Biografico degli Italiani*, vol. 14 (1972): 191–194.
74 This, for instance, was the case for the cardinal of Bologna Carlo Oppizzoni. From 1796, he enjoyed the advantages of being an early supporter of the pro-French governments, becoming 'one of Napoleon's most favoured prelates'.

However, after the authorities were unable (or unwilling) to cover up a sexual scandal that involved him in 1806, he became one of the Italian government's harshest opponents. Ultimately, he became one of the thirteen "black cardinals" jailed by the Napoleonic authorities until 1814: Ambrogio A. Caiani, "Collaborators, Collaboration and the Problems of Empire in Napoleonic Italy, the Oppizzoni Affair, 1805–1807," *The Historical Journal* 60/2 (2017): 385–407.

75 Plea by the priest of Belvedere to the vice-prefect of Portogruaro (2 February 1811), ASVe, San Donà, b. 80.

76 It was the mayor of Grado himself who revealed this practice, following the initiative of the local justice of the peace. Apparently, no legal action was taken against the municipal registrar: letter from the mayor of Grado to the vice-prefect of Portogruaro (20 October 1812, n. 1852), *ibid.*

77 Copy of a letter from the priest of Belvedere to the mayor of Grado (6 July 1812, n. 22), *ibid.*

78 Copy of a letter from the priest of Belvedere to the vice-prefect of Portogruaro (28 October 1812), *ibid.*

79 Letter from the parish priest of Greghentino to the interior minister (25 August 1806), ASMI, Popolazione p.m., b. 4.

80 Letter from the parish priest of Bernareggio to the vice-prefect of Monza (21 October 1811), quoted in *Arti e tradizioni popolari sui costumi e le tradizioni nel Regno Italico*, ed. Giovanni Tassoni (Bellinzona: La Vesconta, 1973), 109–111.

81 Letter from the mayor of Meolo to the prefect of the Adriatico (12 June 1809, n. 214), ASVe, Prefettura, b. 226.

82 List of possible civil registrars in the annexed territories of the department of the Adige (undated, but March 1807), ASMi, Popolazione p.m., b. 69.

83 'Elenco degli individui che hanno eseguite le funzioni d'ufficiale, od aggiunti ufficiali dello Stato Civile negli anni 1806, 1807 nelle comuni del dipartimento dell'Adige' (undated), ASMi, Popolazione p.m., b. 5.

84 Antonio Lazzarini, "Problemi d'impianto dei servizi demografici in un'area della montagna veneta: il dipartimento della Piave," *Archivio storico di Belluno, Feltre e Cadore* 299 (1997): 98–113.

85 Letter from the prefect of the Piave to the interior minister (25 December 1806, n. 4628), ASMi, Popolazione p.m., b. 80.

86 Letter from the prefect of the Piave to the interior minister (21 February 1810, n. 1160), *ibid.*

87 Copy of a letter from the mayor and registrar of Bottrighe to the district's vice-prefect (27 December 1808, n. 3), ASVe, Prefettura, b. 226.

88 Letter from the mayor of Aquileia to the vice-prefect of Portogruaro (14 February 1811, n. 176–177), ASVe, San Donà, b. 56.

89 Letter from the registrar of Portogruaro to the vice-prefect if Portogruaro (18 January 1812, n. 60), ASVe, San Donà, b. 80.

90 Letter from the mayor of Fossalta di Piave to the vice-prefect of Portogruaro (10 June 1812, n. 671), ASVe, San Donà, b. 56.

91 Letter from the mayor of San Stin di Livenza to the vice-prefect of the IV district (1 August 1812, n. 1807), *ibid.*

92 Letter from the registrar of Torcello to the prefect of Adriatico (24 November 1806, n. 16), ASVe, Prefettura, b. 90.

93 Letter from the registrar of Torcello to the prefect of Adriatico (26 January 1807, n. 24), *ibid.*

94 Letter from the mayor of Caorle to the vice-prefect of Portogruaro (17 April 1811; n. 433), ASVe, San Donà, b. 56.

95 Letter from the mayor of Caorle to the vice-prefect of Portogruaro (24 May 1811, n. 519); letter from the mayor of Caorle to the vice-prefect of Portogruaro

(31 May 1811, n. 538); letter from the mayor of Caorle to the vice-prefect of Portogruaro (11 June 1811, n. 579), all in *ibid.*

96 Draft letter from the vice-prefect of Portogruaro to the parish priests of Cottoni and Zorzi (30 April 1811, n. 2586), *ibid.*

97 Letter from the prefect of the Olona to the interior minister (27 September 1806, n. 10971), ASMi, Popolazione p.m., b. 4.

98 Bovara's circular mentioned that the clergymen's involvement could only be on a voluntary basis. The parish priests understood this clarification as an overall prohibition to act as civil registrars: letter from the interior minister to the minister of Worship (16 March 1807, n. II/3371), *ibid.*

99 Letter from the mayor of Pramaggiore to the vice-prefect of Portoguaro (29 December 1811, n. 1021), ASVe, San Donà, b. 56.

100 Letter from the mayor of Cinto to the vice-prefect of Portoguaro (16 December 1812, n. 1418), ASVe, San Donà, b. 80.

101 Plea by the priest of Teglio to the vice-prefect of Portogruaro (15 August 1812), *ibid.*

102 Plea from the parish priest of San Stin priest to the prefect of the Adriatico (28 March 1812), *ibid.*

103 After the return of the Napoleonic authorities to the Venetian departments, several parish priests were accused of having had pro-Habsburg behaviour during the occupation: Valentina Dal Cin, *Il mondo nuovo. L'élite veneta fra rivoluzione e restaurazione (1797–1815)* (Venezia: edizioni Ca' Foscari, 2019), 276–282.

104 In some of these cases, however, it was the civil registrar himself who had interrupted his activities: copy of a table made by the secretary of the vice-prefect of Feltre (12 September 1809), attached to the letter from the prosecutor of Feltre to the justice minister (15 September 1809, n. 1312), ASMi, Popolazione p.m., b. 80.

105 Copy of a letter from Giuseppe dalle Nogare to the prosecutor of Schio (undated); declaration by the Salcedo notary Giovanni Poletto (undated), both in ASMi, Popolazione p.m., b. 73.

5 A Situational Use

The Population and the New Identification System

On the morning of 30 June 1809, the Vicenza police asked for the identity document of a poorly dressed young man in the city's main square. The man produced two documents that attested his identity as that of Antonio Bossina from Verona: one document written in German and dated 1800, which he described as a passport, and an Austrian discharge paper written in German and French, issued in Lviv in January 1809. It is not possible to know whether the police agents understood either German or French. In any case, having verified that Bossina did not possess a regular security card, the agents arrested him and took him to the local prison. The following day, Bossina stated that he was a 27-year-old former Austrian soldier from Verona, where his father Valentino was still alive. He had come to Vicenza to look for a job as a baker, his occupation before he had been enlisted nine years earlier. However, when the police commissioner pressed him with more questions, the man declared that Antonio Bossina 'was not his name' and that Verona was not his birthplace. He now presented the police a parish certificate of christening, stating that he was actually Agostino Brazzo from the small village of Limena in the countryside around Padua.

After this revelation, the police commissioner decided to leave him in prison for a few days. During his second interrogation, the man explained the origins of his double identity. He had invented Antonio Bossina when he enlisted in the Habsburg army. At that time, his plan was to desert as soon as he had drawn his first salary as soldier. However, something must have gone wrong: under the name of Antonio Bossina, he had served for the next nine years as a fusilier in a Galician infantry regiment. After being dismissed earlier that year, he had come back to his family, but he had soon realised that he needed to find a job elsewhere. He had then moved to Vicenza, where he had a brother and some contacts. He had not bothered to apply for a security card for the trip because 'everyone [had] told me that these [Austrian] documents were sufficient'. Despite having used his Austrian identity and documents at the time of the arrest, he now wanted 'to be known for whom I truly am': Agostino Brazzo from Limena, a young man looking for a job as baker.[1]

DOI: 10.4324/9781003412434-6

The case of Antonio/Agostino is exemplary in its uniqueness. This man had lived almost one-third of his life under a forged identity. For his brothers in arms, he was undoubtedly Antonio from Verona. However, he never really forgot his original identity. He asked the parish church in his hometown for a certificate of christening in 1804, four years after his recruitment. Still, once he had been dismissed, he kept using his soldier's identity. Even if he was certainly Agostino when he returned to his family for a few days, he soon became Antonio again when he travelled to Vicenza. At first when he was arrested, he continued to perform his "Austrian" identity, but he soon returned to his first identity. In all these changes, he used his paper documents selectively to confirm one or the other of his claimed identities. This kind of fluidity could appear unusual, but it was natural for an identification culture based on performance. He could be both Antonio and Agostino, even if not at the same time. Although he knew of the obligation to carry a security card, he thought that his Austrian and religious documents would be sufficient to guarantee his double identity. Obtaining a security card would have ultimately implied a choice between one of his two identities.

As we will see in this chapter, the behaviour of Antonio/Agostino summarises the way most people behaved when faced with the new state-driven identification system and the understanding of legal identity that it implied. Misunderstandings, conscious misuse, indifference, and situational use were widespread in Napoleonic Italy. Each in their own way, they all demonstrate how the traditional culture of identification was not simply replaced by the new state in the everyday life of Italians.

New Passports, Old Understandings

While the issuing of security cards remained an administrative procedure, the authorities still thought of passports mainly as a political matter. Theoretically, every citizen who wanted to obtain a passport had to apply to the interior minister himself, according to the traditional understanding of passports as a form of privilege granted by the sovereign.[2] In practice, however, the local prefecture awarded passports more often than not, even when the application was not a matter of urgency – which was by law the only condition in which the prefecture could intervene.[3] People thus applied for passports in an older way: through pleas. Pleas – which were traditional tools of communication between rulers and subjects – were therefore integrated into the new identification system managed by the state.[4]

As the Bolognese archives show, the use of pleas was common for ordinary people as well as for important personalities. When the cardinal of city – and senator of the kingdom – Carlo Oppizzoni needed to travel to Paris, he requested a passport by way of a plea, just like any other citizen would.[5] These requests did not follow any precise procedure or require any forms. Similarly, no precise criteria were set for the acceptance of requests. As was the case for any other plea, the response was at the discretion of the authorities, who could freely decide whether to grant the passport or not. The applicant thus needed to include the reasons and guarantees that could help them obtain the document in their plea. The traveller's final destination and the reasons underlying the trip were specified in every request. These were often signed not only by the applicant, but also by one or more guarantors. Besides these common features, the requests varied widely in length and in content. In most cases, the requests were accepted – especially if they included credible guarantors. The foreigner Geneso Visconti, for instance, obtained a passport for travel to Livorno even if he did not have an identity certificate, because he had received a guarantee from a Bolognese citizen.[6]

The pleas also show how the applicants used specific types of arguments to sustain their requests. In June 1809, Stefano Coppari, obtained a passport to return to his hometown of Gualdo in the newly annexed territory of the Marche. In his plea, he highlighted that he had been freed by the French army from a prison in Perugia where he had been jailed as a 'victim of the persecution of the priests and of the now extinct Inquisition'.[7] Political merits were not the only arguments that could be mobilised by applicants. They could also mention capabilities that could be exploited in the future. This was the case for the Florentine Pietro Bertini, who in November 1810 asked the local police for a regular security card. In his three-page request, he underlined that he was in Bologna to make wine 'similar in colour and taste to the Chianti wine of Tuscany' and that he was planning to introduce innovations to Italian wine production. He also promised to solve, in three years at the most, the issues related to a particular parasite that ruined the harvest in previous years.[8] Despite his claimed merits, his request was rejected, and he was only granted a temporary security card for foreigners.[9]

Since the issuing authorities were not required to motivate their decisions, the reasons for refusals often remained unclear. In other cases, however, refusals were quite self-explanatory. This was the case, for example, for Melchiorre Marsigli, a 42-year-old rope maker who requested a passport to Paris for himself and his daughter. Facing this request, the police officer pointed out that Marsigli – commonly known by the nickname 'Bonaparte' – was 'terrible in every respect' and had been investigated several times by the judiciary. He claimed that the state owed him money – and, for this, 'he had continuously annoyed the authorities'. "Bonaparte" apparently wanted to travel to Paris in order to meet the real Napoleon and ask for his money to be paid. The police officer rejected the request because of Marsigli's complete lack of resources and his questionable sanity. He also withdrew his security card.[10] Since suspects could not obtain an ordinary security card, an individual like Marsigli could certainly not be given a passport.

Even if these documents were theoretically individual, requests were often collective. This practice betrays the persistence of a traditional understanding of documents, according to which registration procedures were often carried out only by the heads of families or employers.[11] Collective requests sent to the prefecture of the Reno included, at the bottom of the page, a list of the people the request referred to – along with some information on each of these, such as their place of origin or age. In May 1807, for instance, Severino Sacchetti requested a security card for himself and his servant because he needed to travel across the kingdom, even if he was still of conscription age.[12] The following year, four men – coming from two neighbouring villages in the Apennines – asked for a passport with which to travel to Tuscany for business reasons. They presented their security cards that were withdrawn, and received 'individual passports'.[13] It was often employers who handed in these collective requests. In May 1808, Luigi Bevilacqua asked – via his agent – for security and residence cards for ten relatives and servants that he had moved from Florence to Bologna with him.[14] A similar request was made by the two owners of a glass factory for 14 workers that had already arrived in Bologna from 'the state of Piedmont'.[15] These requests were accepted by the prefect, who simply inspected the certificates with which they had arrived. If the head of a family, or an employer, was reputable in the eyes of the state, this guarantee also extended to other members of their community.

Collective requests could also not be business-related. In March 1807, the 18-year-old Michele Negri asked for a passport for himself and five other inhabitants of his village Argelato to travel to the basilica of Loreto, in the Papal states. They had been randomly selected to be a part of the delegation that their village traditionally sent to the sanctuary with a donation: a 'very ancient pious custom'.[16] A different purpose lay behind a request for security cards submitted by 19 citizens of Bologna. This colourful group of people – composed of young artisans, librarians, students, and shopkeepers – wanted to spend a Monday in March at the castle of Budrio, 'in order to amuse their families'. Since some of them were still in conscription age, while others did not have a security card at all, they feared that they would be arrested by the gendarmerie. Thus, they asked for a security card valid only for one day: 31 March 1807.[17] The prefect instead gave them a different pass (*Non si molesti*) that was not envisaged by the national legislation.[18] The presence of young men devoid of security cards did not seem to trouble him. Furthermore, the list of applicants only contained male names. It is to be assumed, then, that the rest of the families (namely the women and the children) did not need any sort of state certificate to travel to Budrio, since the gendarmerie would have probably not checked their identity – or their conscription status.

The cases demonstrate how traditional practices of identification – such as pleas and collective requests – were still active during the Napoleonic era. The authorities themselves accepted and abetted these understandings of identification in their administrative procedures. The persistence of pleas and collective requests show how old ideas and practices could coexist with new procedures and tools of identification.

Dissimulating Identities

Ordinary people did not always use the state identity documents in the way the authorities imagined they should. The identification system was adapted and undermined through different strategies and practices that exploited its numerous flaws. Indeed, the issuing procedures and the materiality of identity documents did not prevent misuse.

The underlying heterogeneity of practices among different departments resulted in several breaches in the identification procedures. Even inside the same department, clear and uniform procedures were made impossible by the flexibility and the arbitrariness of the bureaucracy. While countless people who did have regular documents were arrested for identification reasons, it was possible, for example, for a Tuscan man to cross the border and travel throughout the kingdom for days with an endorsed *lasciapassare* that was, in fact, a completely irregular handwritten document released by a municipality in the French empire.[19] Any system for the control of identity documents required a large degree of confidence in both the holders and the issuing authorities. Both of these actors often revealed themselves to be quite untrustworthy.

In all the consulted police archives, not a single case of full-blown forgery can be found. A security card could easily have been counterfeited: it did not present any specific forgery-resistant features such as watermarks or complex printed shapes that would have been hard to replicate. On the other hand, the limited spread of printing presses probably limited this practice. To create a series of forged security cards, a potential forger would have needed a combination of equipment and technical know-how that, at the time, was not widespread, and was kept under strict surveillance by the authorities.[20] However, the difference in the layout of security cards from different departments would have made it difficult to identify a forged card. The absence of forged documents in the archives might also be a consequence of their potential undetectability.

Forgery was surely easier for the handwritten elements of the cards. This type of forgery would have been equally hard to detect, since local authorities had the habit of modifying the information contained in the cards with a simple handwritten annotation. This practice surely decreased the cards' effectiveness as identification devices. However, it also left a wide margin of agency both to the general population – who could alter the cards without being detected – and to the police – who could decide whether or not to arrest a suspect with an annotated card, depending on the situation.

This habit created ambiguous situations. In April 1808, for instance, the 19-year-old trader Luigi Longo was arrested with a security card issued by the municipality of Conegliano with the original date ('17 December 1807') crossed out and a new date ('9 December 1807') added on the side. Since the card was valid for six months at most, whether the date of issue had been forged or not was extremely relevant. When interrogated by the police authorities, Longo declared that 'the person who issued the card, having written the wrong date, corrected it, even though it does look forged'. Quite significantly, the police were not able to understand whether the young trader's declaration was true or not, even if they had the security card right in front of them.[21] Documents could also be stolen from local authorities and then compiled at a later moment. In August 1809, for example, a priest ran into 'three unknown men armed with a big stick'. They forced him to sign a piece of paper – which he recognised to be a passport – with the forged signature of a local mayor.[22] Even if it is not clear how they had obtained the blank passport, the thieves needed someone capable of compiling it – and they easily found him.

The frequent annotations made by the authorities also allowed for a large number of potentially forged dates of birth. For instance, the security card of a 21-year-old coachman from Padua contained a simple handwritten annotation of his date of birth, with no marks on it that could demonstrate whether it had been made by a legitimate authority or not: 'Born 6 June 1790, valid until the age of conscription'.[23] The departmental authorities, however, were more worried about the possibility of an applicant asking for a security card with a forged birth certificate. As we have seen in Chapter 4, this certificate was normally issued by the parish priest who kept the religious parish records. In order to prevent counterfeiting, the central authorities made it mandatory for each certificate issued by the parish priest to be certified by the municipality.[24]

The use of security cards originally issued to other people was more frequent. This practice was surely easier – and probably less risky – than producing actual forgeries, thanks to the superficial nature of the information contained in the security cards. If someone was not too dissimilar from the card's physical description, there was no way to detect this type of misappropriation. Security cards belonging to other people could be obtained in several ways. Firstly, social relations were a channel through which identity documents were transmitted. In August 1808, for instance, the gendarmerie arrested a young man at the local festival of the rural village of Montegaldella. Asked about his name and age, the young man declared that he was the 33-year-old Giovanni Battista Bennato and produced a security card that confirmed this identity. The gendarmes nevertheless noticed some differences between the physical description contained in the card and the holder's appearance:

> the aforementioned card – the gendarmes wrote in their arrest report – specifies that the holder is brown-eyed, while, on the contrary, his eyes are grey [and] his age is three years younger.

At that point, Bennato had to declare his "true" identity: his name was Francesco Bennato and he was 22 years old. The card probably belonged to an older family member, whose identity he had appropriated in his declaration.

The gendarmes had become aware of the discrepancy between the document and its owner not because of his age – which they had overestimated by eight years – but because of such a minor detail as the colour of his eyes.[25]

However, identity swaps could also have a commercial nature. Some arrests conducted by the Vicenza police seem to suggest that there existed, in the everyday lives of the Italian population, a proper trade in identity documents, and that this was widespread. In December 1809, a police informer told the Vicenza commissioner that a man from Trieste was in the city, enjoying an unmotivated degree of wealth and holding a security card which he had bought from one of the town's inhabitants. Whether this information had been provided by the seller himself or betrayed by the suspect's lack of discretion is hard to tell, but once he had been arrested, he declared that he had found the security card in the street. A friend of this man's was also arrested on the same occasion, carrying a card belonging to another citizen of Vicenza. When interrogated by the police, said citizen stated that he had lost his card, but it is highly likely that he had sold it himself.[26] Similarly, in January 1810, Francesco Burselli was arrested in the department of the Piave with a security card issued by the department of the Bacchiglione to someone named Gaetano Coletti. Burselli admitted that he had bought this card in Vicenza from 'a carpenter or an innkeeper', probably the Coletti himself – a notorious thief and cheat who had nevertheless obtained a regular security card from the local authorities.[27] Even if we take the explanations provided by the men who were arrested at face value, it is clear that it was relatively easy to obtain a security card that had been issued to someone else.

Identities could also easily be forged in the moment in which documents were issued. Especially in the major centres where interpersonal knowledge was limited, the procedures for the issue of security cards contained no sure-fire way to prevent a card being issued to someone with a different identity. As we have seen, local bureaucrats could – in case of need – resort to witnesses that did not know the applicant. Moreover, two conniving witnesses and a parish certificate (that usually did not contain any physical description of the owner) were all that was needed to obtain a card with a coherent physical description of its owner, but a different name and surname. Such a forgery would have been almost impossible for the authorities to detect. The inefficiency of the issue procedures, however, did emerge from time to time. In May 1808, a vagabond arrested in the rural village of Gazzo declared: 'I had obtained [the security card] in Vicenza through a man whose name and surname I do not remember'. The detainee did not explain this statement and the police did not seem interested in verifying what he was saying.[28] A forged identity was not yet considered a crime that needed prosecuting.

In another case, a forgery at the point of issue was actually implied by a member of the central government. In April 1809, a letter warned the Bologna police that a certain Luigi Compagnoni was arriving from Lugo with an identity document under the false name of "Luigi Modi". The document had been obtained by the man 'with the guarantee of two inexpert witnesses' that had sworn on his (false) identity.[29] As another letter which arrived the same day explained, the man had decided to take a new surname because he was going to Milan to serve his relative Giuseppe Compagnoni, a journalist-turned-politician who was heading the committee charged with drafting the new Italian penal code. His influential relative had asked him to change his surname because of 'his merit and his obtained position'.[30] Despite this second letter that explained the more illustrious Compagnoni's involvement, the prefecture of the Reno ordered that Luigi be arrested as soon as he arrived in Bologna. However, since he was not caught there, he probably arrived in Milan with his new surname – originally certified by two witnesses in his hometown.[31]

From time to time, the police authorities themselves were to blame for the creation of double and uncertain identities. In March 1808, a man was arrested in Vicenza for not having a security card. The man declared that his name was Bortolo Pin, but the gendarme wrote in his report of the arrest that these were 'a supposed name and surname', because the previous year, the same man had been arrested and detained under the name of Pietro Campagnolo. Pin/Campagnolo's interrogation clarified the situation

> I was [under arrest] last year [...] accused of stealing of a *mezzaluna* at the fair in [Grisignano di] Zocco. At that time, having a security card that I had found by chance with the name of Pietro Campagnolo, I called myself by that name in the arrest report: but, in fact, my name is Bortolo Pin.

Even if his real identity had been revealed during the arrest, the man was once again sent to prison and then released under the identity of Pietro Campagnolo. He was probably also sent to his hometown with a *modula IV* under this name.[32]

A mistake on behalf of the issuing authority was also mentioned by the 20-year-old hairdresser Giuseppe Grazioli in May 1808, after he was arrested because his security card did not match with his surname and several features of his physical appearance – the card, for example, did not include the fact that the holder was 'blind in one eye'. In his interrogation, Grazioli explained:

> To have changed my surname, and to have noted that I have a brown beard, were probably innocent mistakes made by the person who issued the card; regarding the omission of my blindness in the distinguishing marks, this can be explained by the fact that the accident to my right eye happened shortly after the card in question was issued.[33]

Similarly, in November 1811, the young peasant Eugenio Fabbro was arrested while trying to cross the border with the French empire in the department of the Reno. He was coming from Gradisca, a small village in the department of the Passariano, and was headed to Rome, where he hoped to find a brother, whose traces had been lost for five years. He had been issued a regular passport for the occasion, but it contained a wrong name: Giorgio instead of Eugenio. During his interrogation, the teenager explained that this was almost certainly a mistake made by the prefecture of the Passariano, since he considered it unlikely that his municipality would issue a certificate with a wrong name on it.[34] The prefect found his declarations to be credible. It was Eugenio himself who had provided a name that did not match the one written in his passport. Moreover, the physical description contained in the passport did match his own appearance. He was therefore released; he regained his passport and continued his trip to Rome, although it is not clear whether the prefecture of the Reno corrected his passport by inserting his "true" name or not.[35] He had travelled more than three hundred kilometres with an identity that was not his – a mistake that was noticed only because he had ultimately declared his "real" name to a patrol.

There was, however, a simpler way to circumvent the identification system implemented by the state: one simply had to avoid police inspections. We have already seen how the location and timing of these inspections were largely predictable in the context of Vicenza. In such an urban context, police checks were more likely to take place at night, targeting young men, unemployed people, and workers. The same predictability in the pattern of police inspections, albeit with different features, can be found at the border crossing with the French Empire of Scaricalasino (literally "unload-the-donkey") in the Tuscan-Emilian Apennines. Here, the guard post was along the main road and document inspections simply happened whenever a traveller crossed the border.

In April 1808, the Tuscan government issued an announcement in which it declared that it would no longer tolerate 'the abuses of several Tuscan cart drivers who [...] enter [the kingdom of Italy] without a passport, or any other identity document'.[36] Despite this announcement, several cart drivers continued to arrive to the Scaricalasino border post without regular passports. The police inspector denied them passage and threatened them with arrest if they tried to cross the border elsewhere. The cart drivers seemed to agree, left their cabs to other colleagues who could continue the journey, and went back to Tuscany. Before long, however, the police inspector found out that the cart drivers,

> instead of crossing the border [at the border post], have crossed over a small mountain and arrived to the road of the Madonna dè Boschi, [where they] wait for their cabs in the tavern.[37]

After his request, effective inspections of the cart drivers' documents were established at the gates of Bologna.[38] In the following months, even if the cart drivers started to take passports with them, they continued to bypass the system of control over their movements and identities whenever they found it convenient. In October 1808, the inspector in Scaricalasino sent a list of 56

Tuscan cart drivers who had not endorsed their passports – as they should have done – in Bologna between 12 and 30 September. Moreover, he had discovered that several of these cart drivers had declared at the border that they had lost their passports, simply because they did not want to pay any stamp duty. The inspector had thus started to ask those cart drivers who had "lost" their passports to also pay the customary endorsement fee. As a result, from that moment on, 'the passports [were] not lost anymore'.[39]

The behaviour displayed by the Tuscan cart drivers is an accurate representation of the ways in which the population could easily circumvent an identification system that the state was unable to enforce, mainly as a consequence of its own shortcomings. Once a new rule was decided on, it was easy to find ways to bypass it. From time to time, the local authorities themselves encouraged this type of behaviour. In December 1811, for instance, the Bolognese police director discovered that several Tuscan cart drivers were crossing the Italian border with French internal passports. He promptly informed the prefect.[40] In his reply, the prefect of the Reno not only candidly justified the fact that this practice had been adopted in the past; he also proposed to exempt the cart drivers, in the future, from the obligation of carrying a passport.[41] The practices adopted by the general population were always related to the local authorities' behaviour, whether these practices were theoretically illicit or not. If the authorities imposed strict controls, people rushed to obtain identity documents. If they did not, people quickly understood and exploited the new spaces in which they could act.

Identity documents could also be used to bypass other laws. In August 1807, the prefecture of the Reno received a collective plea for passports from Maria Bonifazi, a 40-year-old self-defined 'landowner'. The woman wanted to go to the Marche for 'private business' with two maids. One of these was Raffaela Pascoli, an 18-year-old girl.[42] The prefect promptly denied the request. He had been informed that Raffaela was in fact a younger woman by the name of Camilla Pascoli, and that the aim of the trip was to bring her to a convent against her will – a practice which was illegal.[43] Ever since 1805, in fact, the government had forbidden the ordination of nuns who were under 18 years of age.[44]

The real aim of the passport request had been communicated to the prefecture a month earlier by Francesco Boschi, a neighbour who had taken Raffaella's case to heart. After he had made sure that the young girl did not want to become a nun, he had talked to an official at the prefecture, and then sent a letter to the same official specifying the details of the case. Camilla had been forced to take the trip by don Casarini, a priest 'madly fanatic for religion', who in previous years had given the adolescent girl support with the 'diabolical aim' of turning her into a nun. Ferdinando had decided to help the poor girl to avoid the 'regret that I would feel if a teenager were to be unjustly sacrificed'. He also added that she was to be joined by two other girls in her trip to the convent.[45]

After he had denied the passports, the prefect summoned the people mentioned in the request submitted by Marta Bonifazi. Camilla confirmed that she did not want to go to the convent of Cagli, where her sister Raffaela had already been ordained as a nun. She did not have a vocation, and she had repeatedly explained this to both don Casarini and her confessor. She had finally agreed to join the convent only because she was afraid of losing the economic support the priest provided her with, and without which she would have had no means of subsistence.[46] On the same day, one of the two guarantors who had signed the original request for the passports was questioned. Severino Monti Casignoli had never met Camilla personally but seemed to know both the applicant and the priest. He also knew that the aim of the trip was to bring Camilla to the Cagli convent, where she was going to receive a monthly stipend of 28 *paoli*. The fact that Camilla's name had been changed to Raffaella was, in his opinion, a simple 'mistake'. The priest – the guarantor added – could no longer provide for Camilla and had therefore devised this new solution.[47]

Finally, the prefecture summoned Maria Bonifazi, who had applied for the collective passport in the first place. She candidly admitted that she knew that the aim of the trip was to bring Camilla to the convent in Cagli but said that she also hoped that the change of air occasioned by the trip might improve her health. Neither she nor the officer who questioned her at the prefecture mentioned the fact that she had lied about the aim of the trip in her request.[48] Once the investigation was over, the prefect made his decision. The passport was denied to Camilla but was nonetheless conceded to Maria and the other woman who was mentioned as a servant – coincidentally, also a servant of don Casarini's. The priest was to be reprimanded with 'the due admonitions'. No other measures were enforced against the priest or the applicant.[49]

This case shows how identification procedures could be exploited for illegal aims. Without the involvement of a worried neighbouring family, Camilla would have been sent to the convent with a fake name and a different age – conveniently, those of her sister Raffaella, an already ordained nun. In this way, she would have become a nun despite the national laws designed to prevent such practices. Don Casarini was able to produce a credible application that could divert the attention of the authorities from his true aims, by encouraging his servant to request a collective passport. He found two guarantors – probably two friends of his – who were ready to sign a passport request that was riddled with false information. Thanks to this ruse, he almost managed to get the passports. More surprisingly, the authorities did not take any punitive action against him, the guarantors, or the applicant.

This laxity might be explained by the fact that practices like those we have found in this case were probably fairly widespread. In these cases, horizontal control in the neighbourhood was probably a more effective way of preventing the issue of documents under a false name than the identification procedures themselves.[50]

A Difficult Novelty

As we have seen, the general population elaborated a large array of behaviours to exploit the numerous flaws in the state identification system. But people also largely misunderstood this new identification system. The new regulations were often difficult to understand, owing to their vagueness and complexity. They also entailed understandings of identity and identification that were alien to the popular mentality. The discrepancies between different identification cultures were apparent in the everyday misunderstandings and adaptations of the laws and regulations.

First of all, the new rules could do nothing to formalise marginal identities that enjoyed a high degree of fluidity. In October 1810, a young man was arrested near Vicenza for not having an identity certificate. In his interrogation, the prisoner explained that he had been abandoned at birth, and therefore did not have any kind of certificate that could certify his "real age". He was not sure of his date of birth and mentioned his surname as being '*del Pio Luogo*' ('of the orphanage').[51] In his case, it was impossible to provide him with a certified identity, since the man's specific personal identity was undetermined, or at least only partially certifiable. Similarly, another man who had grown up in an orphanage was arrested in December 1809. When the police officer asked him for his name, the man stated that his name was Bortolo Visentin, but the surname on his security card was Valinari. During his interrogation, the day after his arrest, Bortolo explained his position:

> To tell the truth, in that moment I called myself Bortolo Visentin, but since I come from an orphanage, I do not have a stable surname, [and] I call myself Visentin, since I was born in Vicenza, just like the others from the orphanage do.[52]

Abandoned children, once they had become adults, found it hard to fit into the new system of stable and certified identities that the Italian authorities were trying to impose throughout the national territory.[53]

The population also struggled to understand the link between security cards and age. The provisions on this regard were particularly vague: they mentioned the age from which a man could be considered recruitable, leaving the actual decision on whether someone seemed older than 20 or not to the police forces.[54] This vagueness provoked misunderstandings. In August 1811, the Vicenza police commissioner received a request for a security card from a 13-year-old teenager who wanted to move to Padua. The commissioner refused to issue a certificate for the young man because of his age.[55] Still, the strong link between security cards and conscription confused the population. Some of those who were arrested declared that they thought they were exempt from the identification rules because of their advanced age. For instance, a 55-year-old porter declared that he had not obtained a new security card after he had lost his previous one because he thought that, since he was old, he did not need one anymore.[56] The ambiguous relationship between age and the new identification certificates was confusing for both the youngest and the oldest segments of the Italian population.

There was also a widespread idea that good reputation or an elevated social position was enough to exempt an individual from the obligation to carry identity documents. For example, the cook Francesco Navaglia explained his lack of a security card in the following terms: 'I did not have a security card because I never had the time to obtain it. In any case, I am a gentleman and I have worked at the hospital for three years'.[57] In other words, he believed than being a well-known 'gentleman' would have been a sufficient guarantee of his identity. Two years later, the former nurse Antonio Pizzolato made a similar statement to the police upon arrest:

> Since I have been [working] at the hospitals until a few days ago, and did not need a card there, I did not obtain one, believing that I was already known by everyone, and that I would not be bothered.[58]

Members of the police force also believed that their profession exempted them from the identity regulations. The servant Domenico Messi did not replace his torn security card because he was a member of the local national guard and had therefore considered it 'unnecessary' to obtain a new one.[59] For the same reason, his comrade in arms Mattio Rizzo did not apply for a new security card after he had lost his previous one.[60]

Similar attitudes, even if in very different situations, were displayed by French officials during document inspections in the Kingdom of Italy. An imperial employee of the *administrations des droits réunis* of the department of the Arno wrote in July 1808 to the prefect of the Reno to complain about the treatment that he had received during a police inspection at the gates of Bologna. On that same morning, he had been blocked at the gates of the city because he had not endorsed his passport at the local police office. To his surprise, the guards had asked him to get out of his coach – while a porter employed at the gate had shown his 'joy' about the incident. The French functionary had finally managed to get through the checkpoint, but only after he had shown his orders from the French government and surrendered his passports to the guards. Above all, he left with the feeling that he had been treated disrespectfully.[61]

The reactions of French officials to police inspections in Italy could also be of a more dramatic nature. In March 1807, a French cavalry captain trying to cross the border at Scaricalasino refused to show the required documents to the border guard. The official – who could also boast a *Légion d'honneur* – told the border policeman 'with extreme arrogance' that he was not compelled to show any documents. The Italian border guard explained to the French official that he had to inspect the documents of every person who entered or left the kingdom, even if they were dressed in a uniform: if he did not, it would have been easy for criminals with a uniform to cross the border. The French captain, in turn, began to insult the border guard and the Italian government, becoming increasingly violent. Finally, he assaulted the guard with his sabre, and only the presence of other members of the police prevented him from wounding the border officer.[62] The request to provide identity documents was still perceived as a lack of respect, especially for those individuals who thought that their social position exempted them from any form of identity inspection.[63]

Another cause of confusion for the population was the temporary nature of security cards. As we have seen, the identity documents' period of validity was set by departmental provisions and not by the national legislation. In the department of the Bacchiglione, the prefect decreed a general annulment of all security cards issued in the department three times in the first three years after these were introduced.[64] This unpredictable flexibility of the expiration date of security cards was difficult for people to understand. Therefore, it is not surprising that there was a widespread conviction that security cards had no expiration date. In May 1810, two unemployed weavers were arrested in the Vicenza countryside as they were moving to a small town in search of work. They both had an expired security card, one issued in 1806 and the other in 1807. One of the two arrested men declared that, since it had been impossible for him to obtain a new security card because of unspecified issues related to certificates and witnesses, he had decided to 'risk' the journey 'without fear'. He never thought that he could have been arrested, even with his 'old' security card.[65] In March 1808, several men were arrested in an inn for having expired security cards. When asked why he still had security card issued in 1806, even after the cards issued in this year had been annulled by the prefect, the butcher Francesco Zannoni answered: 'I thought that this [card] was valid, and therefore I did not try to obtain a new one'.[66] In July 1812, Domenico Roveda was arrested with a security card issued in 1806, which had thus expired no fewer than six years earlier.[67] The general concept of an expiration date was probably not so easy to understand for people who were used to certifying their identity with documents such as religious certificates that never expired.

If it was difficult to understand that a document could expire, the same was true for the rights of movement implied by each document. In May 1807, two Mantuan clergymen arrived in Bologna. They were headed to Rome, to attend the 'sanctification of five beatified people'. They had been assured in their hometown that security cards were enough to take them to Rome. Therefore, they were quite surprised to discover that they needed a passport to cross the Italian-French border.[68] The following year, two young men were arrested near the border village of Scaricalasino, because they wanted to cross the same border with only their security cards.[69] The two young men were coming from Gradisca, a village in the department of the Passariano. Their destination was Rome, where one of the two arrested teenagers wanted to persuade his father to come back to his hometown, from whence he had departed eight years earlier. During their interrogation, they explained that they believed the security cards were enough to reach Rome, which they thought to be a part of the Kingdom of Italy. This belief stemmed from the teenagers' misconceptions regarding the political borders of the state. But it had also been confirmed by two Bolognese porters, who had told the two boys that they could travel to Rome with a security card – probably to avoid losing two clients on the trip from Bologna to the border. Even if they had been accompanied all the way from Friuli by two inhabitants of a neighbouring village who did have regular passports, the two teenagers seemed to be quite confident that a passport was not necessary to reach Rome.[70]

These cases were not rare. In May 1808, the Scaricalasino police inspector wrote to the prefect of the Reno: 'Some individuals from our kingdom frequently arrive at this post headed to Tuscany with only a security card. For this reason, there are harsh arguments when I deny them exit'.[71] The distinction between different identity documents still had to be understood by part of the population. This was not just because not everyone was aware of the Kingdom of Italy's actual borders, but also because it was difficult to distinguish between the different purposes of the various identification devices. Both passports and security cards were issued by political authorities; they were needed for travel, and they both included personal and physical information on the holders. It is thus understandable that they were difficult to distinguish.

The use of security cards was related to specific situations. Far from embracing the universalistic understanding behind the state's creation of these documents, people's choice of whether to use the state certificates or not depended on the likelihood of being inspected. In the Vicenza police reports, several arrested people declared that they did not have a valid certificate because they did not think they would have needed it. In the words of Santo Cavalli: 'I never leave the municipality, and I therefore never tried to obtain any [documents]'.[72] As we have seen, most police inspections took place at night. It is therefore not surprising that several arrested people declared that they did not have a card because they never went out after sunset. In December 1807, for instance, the sand digger Domenico Zanetti explained that he had an expired security card at home and that he had not applied for a new, valid one because he was 'never out of the house during the night [and] thus [he] believed that obtaining a new one would have been useless'.[73] When he was arrested in November 1811, the porter Lorenzo Moretti justified himself with a similar explanation, even if it was more related to his own personal habits:

> Since I am always busy in the San Biagio barracks serving the conscripts, I never leave the barracks, and it was a coincidence that last night I was at the inn, to which I had come as part of my aforementioned job, since I was there to buy some bread.[74]

Similarly, some individuals decided to postpone their duties of registration and identification due to matters of convenience. In January 1808, a carpenter from Ferrara arrived in Bologna to look for a job. Here, he was lucky enough to find an employer the following day. However, he did not register his presence with the local police because 'it was a holiday' (a Sunday), postponing his registration to the following day. That same night he was arrested by the police, who withdrew his documents and ordered him to leave Bologna. Despite a plea to the prefect, he was effectively expelled the following day – the same day in which he had planned to go to the police to register and apply for the new documents he needed to stay in the city.[75]

The uptake of documents differed between the cities and the countryside. The reduced police presence in small towns implied fewer inspections of identity documents and, consequently, less of a need to have a valid security card. For instance, in 1808, the shepherd Simon Pasquin explained his lack of a security card in the following terms: 'I never have cared to obtain it, since I hardly ever leave the village, and there is not so much strictness in this regard in this area'. When he was arrested, he was wandering the countryside near his place of work, which was coincidentally quite far from his birthplace.[76] Similarly, the beggar Giacomo Marchetti was arrested by the gendarmerie in the village of Sala in August 1807 because he had no identity documents. In his interrogation, Marchetti explained that he lived in Bologna (more than 20 kilometres from Sala), but that he had often left the city in the previous two months to beg in the countryside. Before he was reduced to begging, he had worked as an envoy for several local authorities for ten months. In this period – he explained – he had never obtained a security card, even if his occupations (envoy and beggar) implied a high degree of mobility. Nevertheless, he had never been stopped by the police and had moved across the Bolognese countryside undisturbed.[77] The fact that the use of security cards was not widespread in the countryside could also be due to other reasons. In April 1810, the peasant Valentin Trattenero came to Vicenza to bring some money to his cousin in prison. He did not have a security card with him because it was too difficult to obtain one in his municipality, a small village in the mountains near Schio.[78] The weakness of municipal bureaucracy in the countryside also contributed to the scarce adoption of security cards in rural areas.

Some people coming from the countryside even stated that they were not aware of the requirement to carry an identity document with them. This was the case for a father and son who were arrested for a not having security cards in Vicenza in February 1808. They were living in the countryside near the city, working as hired hands on a farm.[79] Some months later, the 20-year-old cowherd Serafin Crivellaro from the rural village of San Pietro in Gu explained to the police that he was arrested by a police patrol for not having a security card that he did not know he needed.[80] While he was probably aware of the general requirement of having an identity card, he considered himself exempt because he had not been enlisted. In September of the same year, the police arrested a pilgrim who had come to Vicenza from the mountain village of Asiago without a security card. Even if he had lost his card two months earlier, he had not bothered to obtain a new identity certificate for this day trip to the departmental capital.[81] In April 1812, the 34-year-old peasant Bonifacio Comerlato was arrested for not having identity documents while on his way to Cittadella. He declared to the gendarmes: 'I have never had cards ever since I was born, as I have always wandered these territories without cards'.[82]

Similarly, in June 1811, the 25-year-old waiter Antonio Ferrari declared that he had never obtained a security card, even if he should have done so at least five years previously. Three years earlier, he had left Vicenza to find a job in Cittadella, where he had lived for two full years. During the previous five years, he had thus lived in and moved through another department without having a security card, even if he was draftable.[83] Ferrari was living proof that individuals in Napoleonic Italy could live for years without incurring any trouble for not abiding by the new identification regulations. The state was simply incapable of enforcing its own laws. Indifference, adaptations, and misunderstandings were the logical consequence of this incapability.

A Gradual Indifference

If the rural population largely neglected the new identification duties, the situation did not seem much better in the cities. As we have seen, the local authorities in Vicenza exploited the lack of state documents as a tool to control the subaltern and young population. However, the repressive use of the new legislation did not encourage a stable respect of the law in the town's citizens. After a first moment of impact, the urban population also realised that the civil authorities did not have the means to carry out the duties they had set themselves.

When security cards were established in October 1806, the announcement provoked a remarkable reaction by the population. One month after the first announcement, the prefect was forced to postpone the final implementation of the security card system to December, because of the 'large turnout of people who have come in the last days before the deadline'.[84] This huge turnout is also described in the chronicles of a Vicenza nobleman who lived near the prefecture building:

> People come in droves to the Capitaniato [the seat of the prefecture] in order to obtain the safe-conduct card. It is necessary to wait, and now a large amount of people want the card since it must be obtained by the 15th of this month; each one will have to come personally with two witnesses and undergo an examination of several of their features. All the people are registered in two separated sheets of paper and have to sign these sheets of paper together with the witnesses. Five minutes are not enough for one person. The chaos and the crowd are great and remind me the confusion caused by the Austrians during the famous days of the census.[85]

In these first days in which the cards were issued, the confusion was such that some people quoted it as the reason for their lack of a security card:

'I came several times to apply for it – Alberto Negrin remembered in April 1808 – but in a moment of high demand, so I could not obtain it'.[86]

The serial numbers of the available security cards seem to confirm the large turnout described in these chronicles. Between October 1806 and January 1807, the prefecture issued more than 16,000 security cards.[87] Since the area under its jurisdiction numbered more than 54,000 inhabitants, this first issue could be considered to be quite successful.[88] A large part of the 54,000 inhabitants was in fact exempt from the security card legislation because of their age or gender. We can therefore assume that most adult males in the Vicenza district obtained a security card in 1806.

The following periods of issue did not enjoy the same success. As mentioned above, all the security cards in the department were annulled by the authorities three times in the following three years. After these annulments, the turnout was surely not comparable to that which had accompanied the first introduction of the cards. As some annotations by the Vicenza police commissioner show, only 2,078 security cards were issued in the Vicenza district between September 1807 and March 1808. The slow pace of issue was also confirmed during two of the summer months of 1808 (Table VI). This pace decreased even more in subsequent years. Between March and December 1810, for instance, only 983 regular security cards were issued by the Vicenza police commissioner (Table VII).

These numbers reveal that most of the population of the central canton of the department of the Bacchiglione did not obtain new and valid security cards after the first period of issue. This conclusion is confirmed by a bitter observation made by the Vicenza police commissioner in a letter he sent to the prefect in December 1807:

> I must tell you that the population of this municipality is very disobedient towards orders from above and does not bother to get the appropriate card, so that we would probably need ten dedicated prisons if we wanted to arrest all the individuals who do not have a card.[89]

Another confirmation of the increasing public indifference to the laws on identification comes from the arrest reports of the Vicenza police. Several arrested individuals declared that they had lost their security cards and that they had not obtained new ones afterwards. The 45-year-old furnace operator Bortolo Casotto, for instance, provided this excuse when he was arrested by the police for not having documents on a night in December 1807.[91] Similarly, the 39-year-old peasant Domenico Sorio came up with a similar explanation when he was caught in an inn in Vicenza in June 1811.[92] The fact that the police were unable to enforce the laws on identification is also demonstrated by the statements, provided by some of those who were arrested, that they had simply left their certificates at home. This was the case, for example, for the 29-year-old Antonio Tessari, who was arrested at another inn during an inspection by the financial police.[93] Since it was unlikely that someone's

Table VI Security cards issued monthly by the Vicenza police commissioner[90]

12/1807	1/1808	2/1808	3/1808	6/1808	7/1808
854	794	276	154	178	116

Table VII Security cards issued monthly by the Vicenza police commissioner in 1810

3/1810	4/1810	5/1810	6/1810	7/1810	8/1810	9/1810	10/1810	11/1810	12/1810
76	128	132	118	100	97	100	52	90	50

identity documents would be checked during their day-to-day life, it was only natural for many people to leave their security cards at home.

After the first round of issue, people started to disregard the security cards' mandatory nature. This attitude was probably due to an understanding that the police forces' ability to check identity documents was quite limited and tied to certain locations and times of day. Since these documents mainly had repressive and surveillance-related aims, they were only necessary if the authorities decided to inspect them. Their universality remained, in practice, unfulfilled. People mostly chose to ignore the identification duties imposed by authorities who were not able to enforce them. In the space of a few years, the majority of the population ended up largely ignoring the new identification system.

However, the low uptake of state documents does not mean that Italian citizens never used any other means to certify their identity. People continued to use old identification devices and practices instead of the ones that had been recently introduced. The widespread use of different types of identity documents shows how people still acted within the pluralist framework of the traditional identification system.

Several examples of this behaviour can be found in the Vicenza police archives. When, in April 1810, the gendarmerie arrested the wandering cobbler Bortolo Ciolini, he was in possession of a religious certificate and four other identification documents but, according to the police force, none of these certified his conduct.[94] When he was arrested in June 1811, the worker Giuseppe Venzo had two parish certificates of baptism and good behaviour, along with a municipal certificate of exemption from conscription, but no security card.[95] It was not uncommon for employers to issue identity certificates too. For instance, in 1808, an employee of a Vicenza hospital was arrested for not having a security card. He had not obtained one because he had 'a certificate from the director [...] that certified my behaviour'.[96]

The apparent connection between security cards and conscription also led some people to think that military discharge papers constituted enough proof of identity for them to travel throughout the kingdom. When the young Venetian Girolamo Ferro was arrested in Vicenza in July 1809, he declared that he did not have a security card because he thought that his discharge papers 'could have been a sufficient guarantee, having [myself] come all the way here without being obstructed by the armed forces'.[97] Parish certificates could also be enriched with notes relating to the holder's employment situation. For instance, a certificate of good behaviour issued in 1807 by a parish priest and validated by the Vicenza municipality was also annotated by a French commander: 'The war commissioner of Vicenza certifies that the aforementioned [person] is employed at the Santa Maria Nova hospital. The present [document] must serve him as a security card'.[98] Traditional identification practices could be integrated – at least in the eyes of this French official – with the new devices promoted by the state.

However, religious certificates, or *fedi*, remained the most widespread means of identification. These certificates were issued by authorities that could, in theory, certify the behaviour and other personal details of an applicant. However, they never included a physical description of the holder – making them particularly ineffective for identification purposes. These certificates were ultimately not meant to certify the identity of the holders, but rather single aspects of their lives (primarily their behaviour and date of birth). In this sense, they differed quite a lot from the identity documents promoted by the state. Unlike the *fedi*, security cards and passports provided a strong connection between (stable) identities and documents. This connection, however, was often shunned by the population, as it continued to primarily certify its identity with traditional certificates.

*

Far from the debates and legislative reforms which were taking place in Milan, the population reacted to the arrival of the new identification system in a differentiated way. At first, (male) citizens were quick to adopt the new identity documents provided by the state, at least in urban areas. The effective result of the first announcements, however, was quickly jeopardised by a general awareness that the state agents themselves were unable to enforce the new laws. Gradually, people understood that the new identification duties could be ignored and bypassed when needed. Especially in the rural areas, the new identity documents were met with indifference. Since there was no actual risk of being checked in day-to-day life, their mandatory nature remained mainly theoretical. Meanwhile, old means of identification such as the *fedi* stayed in use throughout the kingdom. Some of the local authorities themselves chose to integrate traditional understandings and practices of identification with the new procedures.

The new identification devices, then, overlapped with the old ones without replacing them. People used the new documents only when they were needed – though Italians did not always know when exactly this was. The traditional culture of identification caused substantial misunderstandings of the regulations that made their application even less effective. The deep flaws in the procedures of identification and registration, coupled with the blunders of the issuing authorities, created the conditions for misuse and numerous deceptions. The overall behaviour of the population towards the new identification system was directly influenced by the level of presence that state authorities could exercise in the situations people encountered in their daily lives. Being in a city or in the countryside, being old or young, being out after dark or not: these particular situations directly influenced people's behaviour. For common citizens, the state identification system implied the introduction of new procedures and devices that they could choose to ignore, misunderstand, adapt, or misuse – depending on the situation.

Notes

1 Antonio/Agostino was released after two weeks of detention with a *modula VII*. The police had previously verified his good behaviour with his brother and his formers employers. ASVi, Commissariato, b. 11, f. 1451.
2 Groebner, *Who Are You?*, 171–181.
3 'Regolamento per la polizia generale' (27 March 1802), in *Bollettino*, 1802, 41–47. The 1811 reform certified this practice assigning the task of issuing external passports to the prefectures, while the vice-prefectures could also issue internal passports: Decree (11 June 1811), in *Bollettino*, 1811a, 569–585.
4 On pleas in early modern Europe: Simona Cerutti and Massimo Vallerani "Suppliques. Lois et cas dans la normativité de l'époque moderne. Introduction," *L'atelier du Centre de recherches historiques* 13 (2015).
5 The cardinal obtained the passport, however, alongside a guard of honour for the first part of his trip: letter from the cardinal of Bologna to the prefect of the Reno (1 December 1809), ASBo, Prefettura, 1809, tit. XX, rub. 10, n. 31180. Carlo Oppizzoni was, at that time, a strong supporter of Napoleonic domination within the Italian clergy: Umberto Mazzone, "Oppizzoni, Carlo," in *Dizionario Biografico degli Italiani*, vol. 79 (2013): 384–388.
6 Plea sent by Geneso Visconti to the prefect of the Reno (undated); note by the prefect of the Reno (15 January 1807), both in ASBo, Prefettura, 1807, tit. XX, rub. 10, n. 1329.
7 His plea was accompanied by two guarantees: plea sent by Stefano Coppari to the prefect of the Reno (24 June 1809), ASBo, Prefettura, 1809, tit. XX, rub. 10, n. 16348.
8 Letter from Pietro Bertini to the police director (18 November 1810), ASBo, Prefettura, 1811, tit. XX, rub. 11, n. 242.
9 Letter from the general police director to the prefect of the Reno (31 December 1810, n. 11509), ASBo, Prefettura, 1811, tit. XX, rub. 11, n. 242.
10 Letter from the police officer to the prefect of the Reno (7 March 1808). Marsigli was sent to Budrio, his hometown, with a *modula IV*: note by the prefect of the Reno (8 March 1808), both in ASBo, Prefettura, 1808, tit. XX, rub. 10, n. 4514.
11 This, for instance, was the case in Venice during the last decades of the eighteenth century. Here, the lists of foreigners present in the city only generically mentioned the presence of women, children, or servants: Bertrand, "L'administration vénitienne", 114–115.
12 Sacchetti also mentioned two guarantors for him and his servant. Letter from the Bologna municipality to the prefect of the Reno (15 May 1807, n. 2769/2770), ASBo, Prefettura, 1807, tit. XX, rub. 10, n. 11435.
13 Collective request to the prefect of the Reno (undated, but 04/1808), ASBo, Prefettura, 1808, tit. XX, rub. 10, n. 9388.
14 Letter from Feliciano Martelli to the prefect of the Reno (11 May 1808), ASBo, 1808, tit. XX, rub. 11, n. 9583.
15 Ten of the workers were part of the Bormioli family. Plea sent by Landi and Pisi to the prefect of the Reno (undated), ASBo, Prefettura, 1811, tit. XX, rub. 10 (parte 2), n. 27328. This family was part of the traditional glass production industry of the community of Altare near Savona. In this regard, see Corine Maitte, *Les chemins de verre. Les migrations des verriers d'Altare et de Venise (XVIe–XIXe siècles)* (Rennes: Presses Universitaries de Rennes, 2009).
16 The request also presented two guarantors. Letter from Michele Negri to the prefect of the Reno (undated, but 03/1807), ASBo, 1807, tit. XX, rub. 10, n. 10094.
17 Collective request to the prefect of the Reno (undated, but 03/1807), ASBo, Prefettura, 1807, tit. XX, rub. 11, n. 6228.

18 Note by the prefect of the Reno (27 March 1807), ASBo, Prefettura, 1807, tit. XX, rub. 11, n. 6228.
19 This handwritten *lasciapassare*, modelled on the French passport, was only meant 'for travel in different places of Tuscany'. The Florence police, nevertheless, endorsed the document and added that it was valid for crossing the border and travelling to Cesena. There, the police allowed the holder to go to Rimini, where the document was endorsed with a permit allowing him to proceed to Bologna. However, once the holder arrived in Bologna, he was arrested and expelled from the kingdom: document of Luigi Gualberetti issued by the maire of San Casciano (10 January 1809, n. 12); note by the prefect of the Reno (13 February 1809), both in ASBo, Prefettura, 1809, tit. XX, rub. 10, n. 3906.
20 Gianluca Albergoni, "La censura in età napoleonica (1802–1814): organizzazione, prerogative e uomini di uno spazio conflittuale," in *Istituzioni e cultura in età napoleonica*, eds. Elena Brambilla, Carlo Capra and Aurora Scotti (Milano: Franco Angeli, 2008), 184–219.
21 Since Longo's case was forwarded to the prefecture, it is not possible to know whether he was being honest or not: ASVi, Commissariato, b. 5, f. 393.
22 Letter from the prefect of the Panaro to the prefect of the Reno (7 September 1810, n. 575/3401), ASBo, Prefettura, 1810, tit. XX, rub. 10 (parte 1), n. 34477.
23 Security card of Lorenzo Benini (29 July 1809), ASVi, Commissariato, b. 22, f. 452.
24 This requirement was abolished by the minister of Worship in March 1810. Circular from the prefect of the Bacchiglione to the mayors of the first district (12 March 1810, n. 5619/1787), Bertoliana, Torre, Proclami, vol. A14.
25 ASVi, Commissariato, b. 7, f. 794. For the quotation: 'Processo verbale del detenuto Gio Batta Benatto N°4' (15 August 1808), *ibid.*
26 ASVi, Commissariato, b. 13, f. 1797.
27 ASVi, Commissariato, b. 15, f. 291.
28 Interrogation of Zuanne Camazzola (18 May 1808), ASVI, Commissariato, b. 6, f. 492.
29 Letter from Giancarlo Biancucci to the Bologna police commissioner (1 April 1809), ASBo, Prefettura, 1809, tit. XX, rub. 11 (parte 1), n. 8645.
30 Letter from Zabbarone to the Bologna police commissioner (1 April 1809), ASBo, Prefettura, 1809, tit. XX, rub. 11 (parte 1), n. 8645. On Compagnoni: Giuseppe Gullino, "Compagnoni, Giuseppe," in *Dizionario Biografico degli Italiani*, vol. 27 (1982): 654–661.
31 Annotations of the prefect of the Reno and a prefectural police officer (2 April 1808), ASBo, Prefettura, 1809, tit. XX, rub. 11 (parte 1), n. 8645.
32 ASVi, Commissariato, b. 5, f. 393. For the quotations: arrest report (28 March 1808); interrogation of Bortolo Pin (30 March 1808), *ibid.*
33 ASVi, Commissariato, b. 5, f. 444. For the quotation: 'Processo verbale sopra l'arresto delli Giuseppe Grazioli, ed. Antonio Schiavon', *ibid.*
34 Interrogation of Eugenio Fabbro (25 November 1808), ASBo, Prefettura, 1808, tit. XX, rub. 10, n. 27938.
35 Note by the prefect of the Reno (27 November 1808), ASBo, Prefettura, 1808, tit. XX, rub. 10, n. 27938.
36 Announcement by the presidency of the Tuscan *Buon Governo* (13 April 1808), ASBo, Prefettura, 1808, tit. XX, rub. 10, n. 7490.
37 Letter from the Scaricalasino police inspector to the prefect of the Reno (5 May 1808, n. 101), ASBo, Prefettura, 1808, tit. XX, rub. 10, n. 9373.
38 Letter from the Scaricalasino police inspector to the prefect of the Reno (11 May 1808, n. 101), ASBo, Prefettura, 1808, tit. XX, rub. 10, n. 9792; letter from the Santo Stefano gate inspector to the prefect of the Reno (14 May 1808), ASBo, Prefettura, 1808, tit. XX, rub. 10, n. 10562.

39 Letter from the Santo Stefano gate inspector to the prefect of the Reno (3 October 1808), ASBo, Prefettura, 1808, tit. XX, rub. 10, n. 23805.
40 Letter from the general director of Police to the prefect of the Reno (24 December 1811, n. 32522), ASBo, Prefettura, 1811, tit. XX, rub. 10 (parte 2), n. 32522. The discovery was made possible by the fact that the prefect of the Reno had written to the director to point out that he had released two cart drivers arrested in Bologna for not having endorsed their passports at the police post at the border-crossing in Loiano. They had crossed the border during a storm, when the police guards did not check the travellers: copy of a letter from the prefect of the Reno to the general director of Police (18 November 1811, n. 31687), ASBo, Prefettura, 1811, tit. XX, rub. 10 (parte 2), n. 32522.
41 Draft of a letter from the prefect of the Reno to the general director of Police (28 December 1811), *ibid.*
42 Plea sent by Maria Bonifacci (undated, but 08/1807), ASBo, Prefettura, 1807, tit. XX, rub. 10, n. 18706.
43 Note by the prefect of the Reno (19 August 1807), *ibid.*
44 'Decreto sull'organizzazione del Clero secolare, regolare, e delle Monache', in *Bollettino*, 1805a, 123–132.
45 Ferdinando was also worried that the priest would have stopped supporting Camilla, had she not gone to the convent. He had obtained information about the origin of her father, in order to verify whether she could ask for support from the Pia Opera dei Vergognosi. Since her father was not from Bologna, she would have needed extra support. He also asked the prefecture to provide her with support. Letter from Ferdinando Boschi to the prefecture of the Reno (28 July 1807), ASBo, Prefettura, 1807, tit. XX, rub. 10, n. 18706.
46 Interrogation of Camilla Pascoli (20 August 1807), *ibid.*
47 Interrogation of Severino Monti Casignoli (21 August 1807), *ibid.*
48 Interrogation of Maria Bonifazi (21 August 1807), *ibid.*
49 Note by the prefect of the Reno (22 August 1807), *ibid.*
50 On the crucial role of vigilance in enabling state control see: Arndt Brendecke, "Attention and Vigilance as Subjects of Historiography. An Introductory Essay," *Storia della Storiografia* 74 (2018): 17–27.
51 ASVi, Commissariato, b. 23, f. 792.
52 Interrogation of Bortolo Valinari (17 December 1809), ASVi, Commissariato, b. 13, f. 1763.
53 On the names and surnames of abandoned children, see Roberto Bizzocchi, *I cognomi degli Italiani. Una storia lunga 1000 anni* (Roma-Bari: Laterza, 2014), 183–187.
54 'Regolamento per la istituzione delle Carte di Sicurezza' (undated), ASVi, Commissariato, b. 1, f. 7.
55 Letter from the Vicenza police commissioner to the prefect of the Bacchiglione (30 August 1811, n. 1129), ASVi, Commissariato, b. 22, f. 693.
56 Interrogation of Bortolo Pochetto (25 June 1810), ASVi, Commissariato, b. 15, f. 358. Other similar declarations: Interrogation of Michele Rocco Berton (29 December 1810), ASVi, Commissariato, b. 18, f. 887; interrogation of Pietro Guerra (31 October 1810), ASVi, Commissariato, b. 28, f. 691. The three arrested men were all in their 50s. It is possible that these declarations referred more to the actual police practices than to the legislation. They did not state whether they considered the security card to be 'not needed' because of the former or because of the latter.
57 Interrogation of Francesco Navaglia (8 April 1808), ASVi, Commissariato, b. 5, f. 340.
58 Interrogation of Antonio Pizzolato (30 October 1810), ASVi, Commissariato, b. 17, f. 724.

59 Interrogation of Domenico Messi (3 March 1810), ASVi, Commissariato, b. 14, f. 125.

60 Interrogation of Mattio Rizzo (3 March 1810), *ibid.*

61 Letter from Marin De Lestre to the Bologna police commissioner (12 July 1808), ASBo, 1808, tit. XX, rub. 10, n. 15989.

62 Letter from the Scaricalasino police inspector to the prefect of the Reno (3 March 1807, n. 75). The prefect of the Reno wrote a letter to the cavalry commander in Bologna to report and criticize the attitude of the captain, explaining that the policeman was just following an order issued by a general of the Italian army that only exempted generals from police checks. A copy of the order was sent to the police inspector in order to avoid any further similar occurrence: annotations by the prefect of the Reno (20 March 1807), both in ASBo, Prefettura, 1807, tit XX, rub. 10, n. 4446.

63 It was often the local authorities themselves who embedded these understandings in their regulations. This was the case, for instance, in Vicenza, where the gate guards did not have to check the identity documents of 'distinguished individuals and military men, especially high-ranking ones'. ASVi, Commissariato, b. 12, f. 1692.

64 In July 1807: announcement by the prefect of the Bacchiglione (16 July 1807, n. 1826), Bertoliana, Torre, Proclami, vol. A12; in September 1808: announcement by the prefect of the Bacchiglione (30 September 1808, n. 18105), Bertoliana, Torre, Proclami, vol. A13; in October 1809: announcement by the prefect of the Bacchiglione (1 October 1808, n. 2948), ASVi, Commissariato, b. 13, f. 1723.

65 Interrogation of Nicola Dal Ferro (1 June 1809), ASVi, Commissariato, b. 15, f. 309.

66 Interrogation of Francesco Zannoni (17 March 1808), ASVi, Commissariato, b. 4, f. 287.

67 ASVi, Commissariato, b. 27, f. 418.

68 They obtained the passports with the guarantee of two Mantuan men living in Bologna: request by Giuseppe Sternini and Arcangelo Manzini (undated, but 05/1807), ASBo, Prefettura, 1807, tit. XX, rub. 10, n. 10080.

69 In the report, the Tuscan territory is still mentioned as part of the 'Etruscan kingdom', even if it had been annexed by the Empire almost a year earlier. Arrest report (20 November 1808), ASBo, Prefettura, 1808, tit. XX, rub. 10, n. 27938.

70 Interrogation record of Antonio Clabassi (25 November 1808); interrogation of Sante Contardo (25 November 1808), both in *ibid.*

71 Letter from the Scaricalasino police inspector to the prefect of the Reno (5 May 1808, n. 1002), ASBo, Prefettura, 1808, tit. XX, rub. 10, n. 9372.

72 'Processo Verbale per l'arresto di Santo Cavalli sospetto Refrattario' (15 June 1811), ASVi, Commissariato, b. 22, f. 45.

73 Interrogation of Domenico Zanetti (7 December 1807), ASVi, Commissariato, b., 1, f. 43.

74 Interrogation of Lorenzo Moretti (30 November 1808), ASVi, Commissariato, b. 9, f. 1073.

75 Plea of Giacomo Dolcetti to the Bologna police director (undated, but 01 February 1808); note by the prefect of the Reno (1 February 1808), both in ASBo, Prefettura, 1808, tit. XX, rub. 11, n. 1914.

76 Interrogation of Simon Pasquin (8 April 1808), ASVi, Commissariato, b. 5, f. 341.

77 Marchetti was also a former policeman and had some weapons with him: interrogation of Giacomo Marchetti by the San Giovanni in Persiceto justice of the peace (25 August 1807, n. 4402); interrogation of Giacomo Marchetti by the police office of the prefecture of the Reno (10 September 1807). After two weeks of detention, he was finally released with an injunction to obtain a security card if he wanted to travel across the kingdom: note by the prefect of the Reno (11 September 1807), all in ASBo, 1807, tit. XX, rub. 11, n. 19995.

78 Interrogation of Valentin Trattenero (21 April 1810), ASVi, Commissariato, b. 15, f. 229.
79 In their interrogations, the two arrested men declared the real cause of their arrest of their previous employer, the landowner and scientist Ambrogio Fusinieri, who was seeking revenge. ASVi, Commissariato, b. 4, f. 227.
80 Interrogation of Serafin Crivellaro (19 July 1808), ASVi, Commissariato, b. 7, f. 695.
81 Interrogation of Giovanni Frigo (11 September 1808), ASVi, Commissariato, b. 8, f. 874.
82 'Processo verbale di arresto' (26 April 1812), ASVi, Commissariato, b. 26, f. 266.
83 'Processo verbale per l'arresto del sospetto refrattario Antonio Ferrari' (18 June 1811), ASVi, Commissariato, b. 22, f. 465. Ferrari was exonerated by the Vicenza municipality: it confirmed that he had not been called up. Letter from the mayor of Vicenza to the Vicenza police commissioner (18 June 1811, n. 2185), *ibid.*
84 Announcement by the prefect of the Bacchiglione (15 November 1806), Bertoliana, Torre, Proclami, vol. A11.
85 Annotation (1 November 1806), in *Arnaldo I Arnaldi Tornieri, Memorie di Vicenza che cominciano dall'anno 1767, 18 giugno e terminano al 1822*, Bertoliana, ms. 3108–3111.
86 Interrogation of Alberto Negrin (5 May 1808), ASVi, Commissariato, b. 5, f. 331. A few days later, the same argument was used by another arrested man: interrogation of Giovanni Pasetto (9 April 1808), ASVi, Commissariato, b. 5, f. 342.
87 Copy of the security card of Felice Cavalli (11 January 1807, n. 16154), ASVi, Commissariato, b. 3, f. 116.
88 The prefecture was in charge of the issue for the first canton of the department of the Bacchiglione. For the number of inhabitants in the canton: *Raccolta delle decisioni, regolamenti, circolari ed istruzioni relative agli attributi proprj delle autorità amministrative nel dipartimento del Bacchiglione. Volume primo* (Vicenza: Da Bartollomeo Paroni Tipografo Dipartimentale, 1809), 169.
89 Letter from the Vicenza police commissioner to the prefect of the Bacchiglione (16 December 1807), ASVi, Commissariato, b. 1, f. 65.
90 ASVi, Commissariato, b. 2, f. 116.
91 ASVi, Commissariato, b. 2, f. 89.
92 ASVi, Commissariato, b. 26, f. 200.
93 ASVi, Commissariato, b. 22, f. 452.
94 Arrest report (15 April 1810, n. 16), ASVi, Commissariato, b. 14, f. 218.
95 ASVi, Commissariato, b. 22, f. 444.
96 Interrogation of Giacomo Pagani (18 January 1808), ASVi, Commissariato, b. 3, f. 167.
97 Interrogation of Girolamo Ferro (4 July 1809), ASVi, Commissariato, b. 12, f. 1504.
98 Parish certificate of good behaviour of Andrea Dezzardo (23 December 1807), ASVi, Commissariato, b. 4, f. 194.

6 An Active Indifference

The People and Civil Registration

At about 9 o'clock in the evening on 29 April 1809, something started to move in the small village of Villa del Bosco, separated from Padua by about 30 kilometres of almost uninterrupted fields. During the day, the rumour that the Habsburg army was on its way had spread in the countryside between Rovigo and Padua. By this time, most of the authorities involved in the local Napoleonic administration had already run away, looking for safety to the west, under the protection of the French and Italian armies. Faced with this power vacuum, an armed group of Villa del Bosco's inhabitants gathered in the village square. Their target was the now-empty municipal building. They forced its door and occupied the municipal seat. They then seized all the documents and papers they could find in the building, amassing them in a pile at the centre of the square. After adding some of the municipality's furniture, they set fire to the heap. In the burning pyre at the centre of the village, the civil registers for the previous two years went up in flames alongside conscription documents, twelve chairs, and the municipal seal.[1]

Just like in Villa del Bosco, dozens of municipal archives went up in smoke throughout the Napoleonic Kingdom of Italy in 1809 – first after the arrival of the Habsburg troops, and then again during the July riots against the new tax on flour milling. Similar bonfires, for instance, were lit in July 1809 in the department of the Tronto, the southernmost of the kingdom's departments.[2] Especially in the Venetian countryside, the burning of municipal archives and documents was widespread. By the end of the anti-tax riots, 56 municipal archives had been burned in the department of the Brenta – which contained Villa del Bosco.[3]

These popular movements have been seen as symbolic of the relationship between the Italian people and the new activism of the revolutionary and Napoleonic administration in their everyday lives.[4] Ordinary Italians – according to this interpretation – remained substantially hostile to the initiatives of the state in the social domain. If this were the case, the burning of civil registers during riots and revolts can be seen as an act of resistance against the new conscription duties. People's actions during social revolts are, in this view, seen as open manifestations of the attitude they had developed

DOI: 10.4324/9781003412434-7

in response to the changes that the Napoleonic government had started to impose on them. An attitude of refusal and resistance.

However, if we shift our focus from the exceptional moments of revolt to ordinary times, the behaviour of the Italian population towards civil registration appears to be much more complex. This is not just because the destruction of civil registers had a minimal practical impact on conscription procedures, since conscription lists were mainly based on parish registers. Italians also engaged in very few acts of resistance against civil registration, if one considers the amplitude of the procedures it required them to engage in. Nonetheless, the lack of open resistance does not automatically imply that people openly accepted the new culture of identity. On the contrary, an in-depth analysis of administrative archives reveals a large variety of behaviours that can be summarised as expressions of "active indifference". However, before analysing people's behaviour, it will be necessary to tackle – and possibly deconstruct – the authorities' attitudes towards them.

Beyond People's "Ignorance"

From the point of view of the central and local authorities, civil registration was constantly undermined by the uncooperative stance of the population. During the decade in which the new legislation was in force, the Italian government regularly found itself having to deal with what it saw as one of the most widespread and endemic problems of the civil registration system: the fact that people often ignored their registration duties, especially in the case of births. When the minister of the Interior communicated the annual number of births to the viceroy in 1811, he pointed out that the data based on the civil registers almost certainly underestimated the real number: 'despite the repeated reminders [...] – the minister bitterly wrote – there are some [parents] who refuse to accomplish this important procedure because of their carelessness or obstinate prejudice [against civil registration]'.[5] Even if state officials considered this phenomenon worrying and ubiquitous, they never tried to ascertain its real extent or its causes. This attitude resulted in a sort of low-intensity "panic", to which the authorities ultimately responded with resignation. Every time a new issue emerged tied to civil registration, they stated their belief that Italian citizens were inherently hostile to the new procedures. And about this, they believed, nothing could be done. In the eyes of the Italian authorities, the people's "ignorance" was self-explanatory and, regretfully, impossible to correct.

Contrary to the rules for marriages and deaths, the civil registration of births was also allowed after an infant's baptism had taken place. However, according to the Civil Code, the registration of a newborn had to take place within three days of their birth.[6] It was precisely the strictness of this rule that made the lack of registration of births such a sensitive issue for the Napoleonic authorities. The Civil Code could not be amended, meaning that the authorities had to be strict in its application – even if part of the population appeared to be unwilling to comply with this particular provision.

Already in December 1806, a few months after the activation of the civil registers, the minister of the Interior had to write a circular dealing with 'the negligence and the ill-will of the parents that avoid the civil registration of their offspring'. The minister indulged in bitter musings on what he considered to be a widespread practice. This, he said, had been caused by three different reasons, which he enumerated and discussed individually: 'ignorance, aversion to the laws, and a will to suppress the civil status of infants'. The ignorance was theoretically being addressed by the fact that the regulations required priests to remind people to carry out civil registration, while the opposition to the law was virtually impossible to tackle. If parents were "against the law", nothing could be done, since it would have been impossible to use 'coercive means'. Finally, the minister pointed out that only the judiciary could confront the explicit intention to 'suppress the civil status of infants'. He thus concluded that any possible action taken to incentivise the registration of infants would be 'damaging, or at least ineffective'. The only possible response to the problem was to wait:

> One can hope that time, by increasingly enlightening the population as to their true interests, will produce better results than those that we will wait for [uselessly] until there is a general understanding of the importance of the registers.

Moreover, in the French Empire, legislators had decided to rule out fines for parents who didn't follow the rules on civil registration because of 'several strong reasons'. The provision in the Italian regulations allowing civil registration to take place even after baptisms had been necessary to avoid 'too much of a clash with the religious principles [of the population]'.[7] One year after the interior minister's circular, however, this latter once again had to reaffirm that it was impossible to register births after the prescribed three-day time limit.[8] Wedged between the inflexibility of the Civil Code and what he perceived as people's negligence regarding the rules, the minister of the Interior could only hope that time would change the minds of Italian citizens. The 'indolence or the guilt of the parents' – as one government official defined the attitude of the population – could only be overcome with the passing of time.[9]

However, at the end of 1808, the central authorities decided that the lack of registration was so widespread that it was no longer possible to ignore it. The minister of the Interior explained to the minister of Worship that the local authorities in two different departments had written to him regarding the fact that infants there were not being registered. He stressed that 'several rough mountaineers' were exposing their children to severe legal consequences 'because of ignorance or negligence'.[10] A few months later, the central authorities were informed of another disturbing case. In the municipality of

Borgomanero, almost twenty births had not been registered in the preceding year. The circular the minister sent to the prefects to address this case highlighted the importance of explaining the usefulness of civil registration, especially to rural populations.[11]

For the Napoleonic authorities, the countryside was the most problematic area as far as the lack of registration of births was concerned. This impression was confirmed by a number of inquests that local authorities conducted regarding the problem. In 1810, the prefect of the Tagliamento wrote to the minister of the Interior to report the results of the comparison he had made 'by chance more than through a special inquiry' between the civil and religious registers in Treviso and Oderzo. In Treviso, the departmental capital, the information recorded in the two registration systems in 1809 more or less overlapped: 94.9 percent of the children christened there were also registered with the civil authorities. In the smaller town of Oderzo, however, the figures were much more worrying. Here, only 43 percent of the local infants who were baptised had also had their births registered in the civil registers. The prefect added that, in these two municipalities, 'the culture and the knowledge of the law is certainly greater than in several other municipalities'. Therefore, he assumed that, in other parts of the department, the lack of registration might have been even more serious. Despite repeated reminders from parish priests and mayors, the prefect estimated that between 1807 – when civil registration had been implemented in the department – and 1809, half of the children born in the area under his jurisdiction had not been registered.[12] These kinds of reports from below played an important role in fomenting the central authorities' "panic" regarding the lack of registration. However, the Italian government never ordered a full investigation on this phenomenon and its causes. For the government, the "ignorance" of its subjects was a sufficient explanation for the problem.

If this "ignorance" was an attitude that could not be cured, any harshness employed in trying to correct it would have been useless. At first, local authorities reacted with a degree of forgiveness to cases that emerged of births not being registered. In August 1809, for instance, the prosecutor-general of Brescia admitted that in previous years, he had often given his approval for late registrations of births that had taken place after the three-day deadline. He knew the provisions of the Civil Code, but read these more as 'advice' than as a strict set or rules. He would not have changed his mind if a local registrar had not pointed out to him that the circular issued by the minister of the Interior in December 1807 explicitly forbade the late registration of births.[13]

On the other hand, local authorities often request authorisation for late registration from the central authorities. In 1808, the prefect of the department of the Bacchiglione, for instance, asked the minister how to behave when faced with the lack of registration of births, which – if not corrected – could lead to 'fatal consequences' for innocent infants and their families.[14] When confronted with 'many and many unregistered births' the vice-prefect of

Campo San Piero had also enquired about the possibility of registering them even after the legal deadline.[15] Similarly, in February 1807, the political and legal authorities of the department of the Adriatico were also faced with the decision of what to do in the case of late registrations. The judge of first instance – who had received an appeal from two Venetian parents asking to register the birth of their son well after it had taken place – washed his hands of the affair and decided to involve the prefect.[16] The latter, in turn, tried to delegate the decision to the municipal officials of Venice, who pointed out that they had no jurisdiction over the matter.[17] These repeated attempts to shift responsibility are an indication of how local authorities were actually inclined towards forgiveness for an attitude that they saw as resulting from the hopeless ignorance of ordinary Italians.

This stance did not alter in any substantial way even after the introduction of the French Penal Code in February 1811. Among the 'crimes and felonies against newborns', article 346 criminalised anyone who was responsible for not registering a birth:

> Anyone who, having witnessed a birth, does not declare it, as prescribed by article 56 of the *Code Napoléon,* within the time limit established by article 55 of the same Code, will be punished with a period of detention of six days to six months, and with a fine of sixteen to three hundred liras.[18]

This provision reaffirmed that the main responsibility for registration lay with a newborn's parents or with any witnesses to their birth. No punishment was envisaged, in cases of missing registrations, for priests or registrars.

When the new Penal Code entered into effect, the minister of Justice wrote to the ministers of the interior and of worship. In his opinion, the establishment of a criminal sanction represented a significant change, particularly considering 'how much – especially in the countryside – the observance of this legal provision has been neglected'.[19] The minister of the Interior thus wrote to the prefects to urge them to spread the news that fines would now be imposed for failures to register births.[20] The authorities felt the need to constantly remind the population of this duty. The latter, however, largely ignored it.

Even after the introduction of the Penal Code, local authorities were indulgent towards parents who did not obey the rules. The very harsh fines prescribed by the Code were often reduced by Italian courts of justice. When Giovanni Battista Dal Pezzo, a peasant, registered the birth of his son at the registry office of the hillside village of Cereda, thirteen days after the event in September 1813, the court in Vicenza only imposed a two-lira fine on him, as well as requiring him to cover court fees. The judges in Vicenza took advantage of a different article in the Penal Code to significantly reduce what could have been a fine of up to 25 lire.[21] These light punishments once again show how the local authorities saw the people's attitude towards civil registration as substantially unrepentant. Even repression was not considered as a viable way to correct the "ignorance" of the population.

The Italians' perceived non-cooperation was also explained by the parish priests' supposed lack of support for civil registration. According to the 1806 regulations, clergymen had to 'warn the father of the newborn, or whoever will be present at the moment of the baptism, of the obligation to present [the newborn] to the civil registrar'.[22] There was a widespread belief among the ranks of Italian authorities that parish priests often did not comply – and that this behaviour was one of the causes of the people's supposed "ignorance". Almost every time the central authorities were informed of problems related to the lack of registration, their proposed solution was to remind the parish priests of their obligations. In February 1810, for instance, the government was informed by the prosecutor-general of Udine that, in the village of Ragogna, newborns were rarely registered by their parents. As a solution, the minister of the Interior suggested to the local prefect that he remind parish priests under his jurisdiction of their duty to urge their parishioners to civilly register their children after baptism.[23]

There was also another reason for which, in the eyes of the authorities, cooperation from the clergy was crucial in preventing lapses in registration. This particular worry adds nuance to the picture the authorities had of the people's attitudes to registration. In May 1809, the prefect of the Bacchiglione sent a circular to the parish priests and mayors of his department to tackle what he saw as a widespread reluctance to register births in the district. The circular began with a standard formula, listing the advantages of civil registration, voicing disapproval for the 'ignorance' of those who did not follow the law on the matter, and stressing the serious consequences that a failure to register one's child could have for the life of such 'innocent' children. Then, in the section addressed to the parish priests, the prefect asked the latter to say the following to fathers of recently baptised infants:

> that the sacred ceremony of the baptismal font is not contaminated in any way by a purely civil discipline, and that [baptism and civil registration] do not have anything in common, or repugnant [to each other].[24]

This unnecessary justification did not escape the attention of the minister of the Interior. He wrote to the prefect that he frankly doubted that anyone could have thought of civil registration in these terms and that, even if this were the case, the 'dignity' of public office should have prevented the prefect from intervening in such a 'frivolous discussion'.[25]

It is impossible to know why the prefect used such a particular formulation to address the problems with the registration of newborns. The circular, nevertheless, shows how local authorities perceived popular behaviour on the matter. The "ignorance" of the people could, in their eyes, take on various irrational forms – even rather inexplicable ones. The authorities believed this to be especially true when religious traditions and beliefs were involved.

Ultimately, the explanation given by the prefect of the Bacchiglione in this circular is probably a far better indication of the way the ruling classes perceived popular mentalities than it is a description of these mentalities themselves. The "ignorance" of the population was not just irredeemable in the short run, but also fundamentally irrational.

Indifferent People

In 1811, the parish priest of the suburb of Romiti near Forlì described the complex social procedures that local couples underwent before marrying. After having settled on the conditions of marriage, a matrimonial mediator gathered the betrothed and their parents to toast to their union. The mediator's formula for the toast was: 'god made them, I matched them, and you will become kin'. They then set a date for a dinner in which the two extended families acknowledged this new kinship. The day after this dinner, the couple went to church and to the civil registry to proceed with the marriage banns. The day of the third religious banns, the two families gathered once again and went from the groom's home to the bride's one – where they had lunch together. At that point, the couple would contract civil marriage and – a few days later – hold a religious marriage ceremony. However, the bride would only move to the groom's house on the Sunday after the religious ceremony. On this occasion, she would also be escorted by her family, who 'celebrated along the road by firing several pistol shots'.[26]

This elaborate ritual shows how important the social events that the Napoleonic state was starting to register were. Local communities in Italy considered births, marriages, and deaths to be major social events. The rituals of civil registration were an addition to an already complicated set of social procedures, and had to find ways to coexist with deeply rooted habits. Despite the Italian authorities' deeply held conviction that there was a widespread lack of interest in civil registration, several reports written by parish priests and local bureaucrats for a national inquiry into the customs of the Italian people painted a rather different picture. According to these reports, collected in 1811, civil and religious forms of registration had become closely intertwined during the previous few years.[27] The apparent contradiction between the authorities' perceptions of the uptake of civil registration and the stories told by witnesses on the ground calls for a more in-depth analysis of the social practices that the population developed to face the new duties they were called to fulfil by the new civil registration system.

Performing such an analysis is not an easy task. As mentioned above, the authorities saw people's behaviour as irrational and therefore as substantially incomprehensible. They were consequently not interested in understanding the causes and the reasons that underlay misappropriations and omissions that they saw as simple violations of the law. The voices of the common people who disregarded or adapted the registration rules are consequently almost lost to us. Nonetheless, the silence in administrative sources regarding the interactions of ordinary Italians with the civil registration system is only an apparent one. If we read between the lines of the panic-stricken documents written by officials, we can see how people reacted to the new obligations of civil registration with a wide array of behaviours that cannot be reduced just to "ignorance".

First of all, we must assess whether the widespread reluctance to engage in registration – as perceived by the authorities – was really as pervasive as official accounts made out. As already mentioned above, the government never carried out a general inquiry of the extent of this phenomenon. The only real data that justify this "panicked" approach came from the town of Oderzo, where – as we have seen – more than half of the new births that took place in 1809 had not been noted in the civil register. However, the information from Oderzo was part of a wider investigation that also included information on the state of registration in the local departmental capital, Treviso. Here, there was an almost complete equivalence between the number of infants who had been baptised and the number of births inscribed in the civil register.[28]

A year later, authorities in the department of the Adriatico found a similar equivalence when collecting information on the city of Venice. When, in February 1811, the minister of the Interior asked for figures on the number of births, marriages, and deaths registered in the city, the prefect of the Adriatico discovered a glaring discrepancy in the Venetian civil registers. The number of deaths was far larger than the number of births. The prefect suspected that this unexpected discrepancy may have been provoked by a failure on behalf of some of the city's inhabitants to register births, and therefore asked the local parish priests to compare their religious registers to the civil ones.[29] The result of the comparison confirmed the prefect's original assumptions: over two hundred infants who had been baptised in Venice were missing from the city's civil registers. The figure represented 6.1 percent of total baptisms in town.[30]

The figures from Venice and Treviso, the two departmental capitals, were relatively similar. In both cases, the number of births that had been noted in the civil registers was very high, at least in statistical terms. To be sure, hundreds of infants who had been christened had not been registered with the civil authorities, but this discrepancy can be partially explained by a difference in the criteria for registering and christening dying infants and by a certain mobility on behalf of parents who chose different places in which to register/christen their children: they did not always register infants in the same municipality in which they had christened them (and vice versa). All things considered, it is highly probable that only a very small amount of these unregistered births was indeed a consequence of an overt will to shun civil registration.

The two inquiries conducted by the authorities in Venice and Treviso might indicate the existence of a strong cleavage between urban and rural centres. It may be that, in the countryside, the new civil registration duties were largely neglected, while people took them much more seriously in departmental capitals.[31] As we have seen, this was the conviction of the civil authorities, who always saw the rural population as more susceptible of participating in the widespread "ignorance" that officials ascribed to the Italian population. The bishops' circulars also often stated that problems with registration were much more common in the countryside than in urban centres.[32]

However, the picture painted by the authorities at the time is probably inaccurate. Even if officials had a strong conviction that the rural population largely ignored its registration duties, they never verified whether this was true or not. The cases that did reach the attention of the central government between 1806 and 1814 are few and far between, considering the scale on which the new civil registration procedures were rolled out. Without a detailed comparison of rural civil and religious registers, it is hard to know if the government was right.

Such a large-scale comparison is unfortunately beyond the scope of this book. However, I have been able to carry out a local analysis of the registers in two randomly chosen municipalities – to verify the extent to which civil registration was actually adopted or not at a local level in certain point of the Kingdom of Italy. The two municipalities are Enego, a village in the mountains of the department of the Bacchiglione, and Cartura, a village in the rural part of the department of the Brenta.[33] Both municipalities were far from their departmental capitals and cut off from the main routes of communication – even if they were not exactly isolated. They never incurred any particular problems in the management of their civil registers. In this sense, these two municipalities were very ordinary. They are thus well-suited as case studies for an investigation into people's behaviours when faced with the new civil registration system.

We will begin by examining the marriage registers of Cartura for the years 1806–1811. As we have seen, the authorities were not very concerned about the civil registration of marriages. The registers in Cartura confirm that this stance was well grounded. The parish registers list 80 marriages for the period under examination, while the civil registers list 84. In most of these cases, the civil and religious marriages took place on the same day, or within a few days of each other. There are only two cases in which the religious ceremony was held before the civil one – in open violation of the rules. On the whole, then, the civil and religious registration of marriages largely went hand in hand in the small village of Cartura.

The same is true for birth registrations. The local parish registers note that 329 christenings took place in the period under examination, while the civil registers list 328 births. That the correspondence between the two registers is almost perfect is also due to the fact that in three cases, the civil registrar clearly listed birth registrations as having happened at an earlier date, while four other births for the period were added in 1835, when the civil and parish registers were compared. Nevertheless, even if we see these seven delayed registrations as a form of resistance to civil registration, these are still a very limited amount of the total (2 percent): a much smaller percentage than that of missing registrations in the cities of Treviso and Venice.

Similar conclusions can be drawn from the registers from Enego for the years 1809–1813. Here, the 428 newborns registered in the civil registers are considerably more than the 407 christenings reported by the parish registers. This difference can be easily explained by the fact that the registrar in Enego maintained a practice of registering stillborn babies that were never christened. A more in-depth comparison of the two registers shows that only four christened children were not registered in the civil registers (0.9 percent of the total), proving the efficiency of the civil registration system in a small mountain municipality such as Enego.

Moreover, one case of delayed registration in Enego gives us more information on the practices of local people. In April 1813, Angela Vettori's mother presented her to the Enego registrar, six days after she was born. The fact that it was her mother who brought her to the registrar is already an indication of the peculiar situation of Angela's family. Her mother claimed that she was not aware of the duty to register births – a rare case of what official saw as people's "ignorance" of the registration regulations. Moreover, she stated that her husband was 'imbecile', while her father-in-law was too old to travel to the registry. The family lived in a village that lay within the municipal boundaries, but that was actually located in a valley 600 metres below Enego. Angela had been christened in another town in the valley, to avoid a trip to the municipal seat along a road that was in 'disastrous conditions'. Despite the several social and geographical factors that placed Angela's family on the fringes of the local community, it is significant that this child was nevertheless registered at the municipal registry. Municipal administrations were indeed able to involve more marginalised elements of the population in the civil registration process, even if this implied delays.

The registers from Enego and Cartura show that the neglect of birth registration was much less widespread than the authorities thought it was. However, if we extrapolate from these local cases that, on average, between 1 and 6 percent of total births went unregistered in the Kingdom of Italy, this still means that, each year, tens of thousands of newborns were not registered by their parents at the Italian civil registries. This is still a large number, for which an explanation must be found. As mentioned above, the authorities never tried to understand the reasons underlying the phenomenon. However, the sources point to some possible explanations.

First, the inconvenience of the registration procedures was one possible reason for people not to carry out the registration of newborns. A few months after the civil registration system was put in place, the registrar of the vast municipality of Torcello tried to explain the lack of registration in his jurisdiction. In his words, his fellow villagers were 'miserable and unfortunate by birth; uncouth and uncultured by education'. The municipality's layout was such that people often had to travel for several kilometres to reach the municipal registry: after such a long trip, they often arrived at the registrar's office after the legal deadline for registration had already expired. The trip was not expensive, but its contained costs were still too much for poorer villagers, especially if they then had to wait in town for the registrar to return to his office.[34] Similarly, in 1810, the mayor of Fossalta explained why it was necessary to have two assistant registrars for two villages that were far from the municipal centre. This was because 'if they will be tolerated no more, no one will come to this registry to register newborns, [...] considering the great distance between the villages and the impracticability of the roads'.[35]

In these two cases, then, local officials cited the large extent of their municipal territories as the main explanation for the lack of registration in their jurisdictions. Although it was apparently only expressed in specific local situations, this explanation can actually be said to apply to a large part of the Kingdom of Italy's territory. A trip to the municipal seat could often take hours, if not an entire day, especially in the largest municipalities. We can probably assume that the poorer and more marginalised strata of the population would choose to shirk their registration duties – the usefulness of which they did not entirely understand – if the alternative meant losing a few hours or an entire day of work, with its attendant income. Omissions in the registration system can thus partially be explained in practical terms.

The failure to register newborns can also be explained by a general difficulty to understand the structures of the new Napoleonic administrative system. In particular, new administrative borders often came into conflict with the religious boundaries that had, for such a long time, regulated people's social life. In the small village of Bottenigo, for example, these conflicting borders caused trouble in the first few months after the establishment of civil registration. Bottenigo was under the religious authority of Mestre, in the department of the Tagliamento, while it was administratively part of the municipality of Torcello, in the department of the Adriatico. When, during the summer of 1806, two inhabitants of Bottenigo died, their deaths were correctly registered at the registry in Torcello, after which they were buried in the cemetery in Mestre. The registrar in Mestre, however, complained to the prefect of the Tagliamento that the deaths should have been registered in his own registers, which were now incomplete. The prefect then wrote to his counterpart in the department of the Adriatico, ostensibly to support the objections coming from Mestre.[36] In the early stages of implementation of the civil registration system, civil authorities also seem to have had trouble understanding the extent of their jurisdictions.

The case of Lastebasse, a border municipality in the Astico valley, however, is an even better representation of how hard it was for people to understand the new administrative divisions. Some small villages in Lastebasse, all of which were about four kilometres away from the municipal seat, were, for religious purposes, part of the parish of the nearby town of Brancafora. The latter town had been part of the Kingdom of Bavaria until 1810, when it had been annexed to the Italian department of the Alto Adige. The inhabitants of these villages had traditionally celebrated and registered their baptisms, marriages, and funerals in Brancafora – across the national and, later, the administrative border.

As the prosecutor of Schio explained in October 1812, this peculiarity had led the local population to persistently neglect its civil registration duties. Ever since 1808, according to the prosecutor, they had not carried out civil registrations of marriages, births, or deaths. They only marked these events with religious ceremonies held by the priest in Brancafora. The previous prosecutor of Schio had tried to intervene to end the practice, but had only enjoyed partial success. Marriages and deaths were now being noted in the civil registers, even if reluctantly, because the parish priest could not celebrate religious ceremonies linked to these two types of occasions without prior proof that they had been registered. Births, however, were still not registered in the municipal registry of Lastebasse.

The prosecutor therefore proposed moving these villages out of the parish of Brancafora, in order to end any 'customary right' that could justify the unlawful behaviour of these 'mountaineers'. He also wrote to the neighbouring vice-prefect to request a detailed list of all the baptisms that had taken place in Brancafora over the previous two years. The prosecutor felt that he needed to give a 'public example' to the local population by punishing those parents who had registered the births of their children in the civil register. Still, he wanted to wait for the result of the inquiry before he acted. He felt that, if too much punishment were meted out, a certain amount of restlessness and disorder might have emerged in the valley.[37] Evidently, any serious repression of the failure to engage in registration would not have been accepted by the local population.

The prosecutor-general of Vicenza supported the prosecutor of Schio's proposal to adapt the religious border to the administrative ones.[38] The minister of Justice, on the other hand, ultimately decided not to endorse this solution, as it probably would have led to an argument with the minister of Worship. Instead, he wrote a letter to the latter asking him to remind the parish priest in Brancafora of his duty to cooperate with the civil authorities.[39]

The case of Lastebasse is significant for several reasons. First of all, it sheds light on the general population's attitude to civil registration. The inhabitants of these remote villages continued to register their vital acts as they had always done, by going to their local parish priest. This attitude came into conflict with the state's new demand for them to carry out a civil form of registration too – a demand whose material usefulness was probably unclear to them. As a consequence of this lack of understanding, they simply continued to not register births, marriages, and deaths in the municipal registry. Since the state did not have the strength to enforce the rules on the matter – and was likely not even interested in doing so – it would be an exaggeration to define this popular attitude as an act of resistance. At first, many people in the valley simply ignored the new civil registration system. In this sense, they really were "ignorant", though not in the same way that the civil authorities imagined they were.

When, after a significant delay, the civil authorities discovered the situation, they had to recur to a representative of religious authority (the parish priest of Brancafora) to obtain a small degree of improvement. When faced with this alliance between civil and religious authorities, the local population changed its ways. People began to register marriages and deaths in the civil registers because they could not do otherwise. However, they continued to ignore the requirement to register births, since this was not a requirement for baptism. Once again, the civil authorities were unable to solve the situation, while there was no need for the population to display any form of active resistance in order to avoid the new registration duties.

It is significant that the prosecutor of Schio only contemplated punishment if the number of infractions was not excessive. After the revolts of 1809, he was probably well aware of the fact that an iron fist would have provoked an active reaction on behalf of the valley's inhabitants. The minister of Justice found that the solution of changing the traditional religious borders would have been even harder to achieve. Ultimately, he responded to an issue that had been caused by the shape of the national borders by recurring to the moral sway of the parish priest – the only authority that, in his eyes, could somehow influence people's attitudes with regard to civil registration. In a case like this, the fact that the construction of the administrative state relied on the clergy is quite evident. There was no need for the valley's population to resort to active resistance, since the civil authorities did not have the power to intervene. Moreover, officials believed that tackling the avoidance of civil registration was probably not worth triggering a potential open revolt that might have had unforeseeable consequences.

In the case of Lastebasse, as in many others, the reluctant attitude of the local population can be summed up as being composed of a mix of indifference and adaptation. People remained indifferent to the new registration duties until they were forced to comply with them. Afterwards, they adapted – but only when civil registration became a necessary prerequisite for the celebration of religious ceremonies. Indifference and adaptation well describe the overall response of the Italian population to the new registration duties. Italians were not "ignorant" – not in the way described by officials. Their reaction to the new civil registration was neither one of open acceptance nor one of open resistance. In the vast majority of cases, they respected the obligations prescribed by the law and registered their newborn children. They did this simply because they were compelled to do so in order to proceed with religious ceremonies – and also because complying with the new obligations did not require too much of an effort.

However, in cases in which it implied a certain burden, or when the authorities did not enforce the rules, compliance with the new registration duties was lax. As noted by the authorities, people were generally indifferent to the supposed advantages of civil registration. However, they were not "ignorant" of these duties, nor of their own interest. Despite officials' efforts to explain why this new form of registration was necessary – alongside older forms of parish registration – Italians did not embrace the civil registration system. In most cases, they adapted to it, although far from enthusiastically. This attitude might explain the authorities' exaggeration of the number of births that went unregistered. They acknowledged the populations' overall indifference, but misread it as "ignorance" – and thus largely overestimated its extent.

Beating the State at Its Own Game

This overall attitude of indifference and adaptation did not mean that people were completely passive with regard to the new registration duties. As a matter of fact, countless ordinary Italians did avoid these duties, or found ways to make the civil registration system serve their own interests.

In most of the cases in which administrative sources detail resistance, this is of a collective nature. Behaviours that contradicted the regulations were often assumed by a large part of a given local community. This is something that may partially explain the authorities' panicking. In most cases, when they received information that a given municipality was experiencing problems with civil registration, these were being caused by a significant part of the local community. This was the case, for instance, in Oderzo, where a substantial number of births were not registered in 1809. However, as the cases of Enego and Cartura show, these forms of collective behaviour only arose in very specific places. What is more, when irregularities in registration were more isolated, they were rarely spotted by the authorities, and thus do not appear in the sources they produced.

One of the strategies that Italians developed to take advantage of the registration system was based on the provision that allowed them to contract civil marriage before the religious ceremony. As the bishop of Concordia explained in 1811, this provision allowed people to bypass the impediments that forbade religious marriages in certain specific situations. On several occasions, the parish priests of his diocese had had to call off marriages whose civil rites had already been celebrated, creating embarrassing situations for 'public morality'. The bishop thus asked the civil authorities to urge local registrars to verify whether couples seeking to get married had already settled on a date for their religious wedding before proceeding with the civil ceremony.[40]

However, the main reason that led people to celebrate civil, and not religious marriages, was to evade conscription.[41] According to the 1802 conscription law, married men were among the last people who could be potentially called up. Consequently, married men were almost certain not to be drafted.[42] This provision gave birth to a plurality of marriage strategies, some of which came to the attention of the central authorities.

In January 1809, the minister of war proposed that men between the ages of 18 and 25 only be allowed to marry after receiving ministerial authorisation. He had noticed that young men often celebrated civil marriages, and not religious ones, to avoid conscription. In some cases, the brides were 'women between the ages of 60 and 80'. In other cases, these married couples continued to live separately after their marriage until they could ask for a divorce.

According to the minister, these practices were 'against the well-being of society and the spirit of the law', whose goal was to promote 'the propagation of the species and the education of children'. Ultimately, they represented a 'constant source of the corruption of customs'.[43]

Similar cases emerged in the following months in certain areas of the kingdom. In March 1809, the mayor of Chioggia reported that, ever since a naval draft had been announced, requests for marriage 'had multiplied [...] in an extraordinary way', and that the local registry could not keep up. He explained that:

> it is well known that this rush to the registrar is not happening because the celebration [of these marriages] is urgent, but rather in order to take advantage of [the] status [of being married] for the draft, since – it is well known – that the church benediction is universally delayed because, according to the religious spirit, this is the definitive act [of marriage], and all this demonstrates that the rush is not for the ultimation of the marriage, but for other aims.[44]

A few months later, the justice of the peace of the town of Porlezza reported that, in his canton, young men had recently been celebrating civil – and not religious – marriages and had been leaving their brides ('innocent deluded damsels') in their family homes. In response, the local prosecutor-general proposed that these young men be located and drafted, as punishment for their behaviour.[45] Similarly, in March 1810, the justice of the peace of Carpi reported that there were at least ten couples in a similar condition in his canton. This situation provoked 'constant unrest, immorality, and depravation' in these local communities. After their civil marriage, these couples continued to live separately. The women involved in these marriages were often, according to justice of peace, 'sacrificed and left to moan in their fathers' houses, and destined to perpetual undesired infertility'. The rare meetings between these married couples provoked 'nothing but altercations, insults, and disorders'. Also in this case, the justice of peace proposed punishing the men who had taken advantage of this loophole in the regulations on marriage to avoid conscription.[46]

It is not surprising that, after such cases came to light, the Italian government discussed changing the 1802 conscription law. In 1810, the minister of war proposed scrapping the provision that gave young married men the right to be placed at the end of the conscription lists, in cases in which their brides were over 60 years old. However, this proposal was rejected by the Council of State, who highlighted that it was based on only two known cases.[47] Faced with the impossibility of solving the problem through legislative initiative, the Italian government resorted – as it had in many other cases – to the clergy. In a circular written in October 1810, the minister of Worship acknowledged that 'these serious and frequent disorders' were caused by the underlying idea that 'the civil register is simply an empty formality'. After their civil marriage, these couples often tried to obtain a religious annulment, but – faced with legal opposition – they often remained separated or were forced to enter into 'unhappy and unfortunate' relationships. The support of the parish priests was thus necessary to defend the 'most sincere rules of private and public morals'.[48] However, the religious authorities' efforts in this sense did not substantially change people's behaviour – at least not in the short run. A few months after the minister's circular, the prefect of the Adige complained about the fact that civil marriages were 'not infrequently' contracted only to avoid conscription. He proposed verifying whether married potential conscripts were actually living with their brides before granting them any privileges.[49]

The cases that we have analysed detail a social practice that had spread to all corners of the kingdom. However, these cases only represented a small proportion of the total number of marriages that took place each year in Italian territory. Once again, the Cartura marriage registers can help us assess how widespread this type of practice actually was. In the small village of Cartura, only 4 of the 84 civil marriages celebrated between 1806 and 1811were not followed by a corresponding religious ceremony. However, there are ten cases in which the civil ceremony took place quite some time before the religious one – with the longest delay being five months. The latter case involved the 'wealthy' 24-year-old Giosafatte Franceschini, who contracted civil marriage with Domenica Leonardi in late June 1809, just after the end of the Habsburg occupation of the area. Franceschini then waited until the end of October to proceed with the religious ceremony: by then, he was probably already 25, and thus no longer in danger of being conscripted.[50] Further information on the practice of marrying to avoid conscription comes from the inspection of the Venice marriage registers that the prefect of the Adriatico carried out in 1810. In the inspection's report, the number of civil marriages (665) celebrated in the city was only slightly higher than that of the religious ones (660).[51] Aside from more exceptional cases, in which an entire community would adopt a particular practice related to registration, these types of behaviour remained mostly isolated – and thus went largely undetected by the authorities.

However, the practice of delaying religious marriage ceremonies could also backfire on those who adopted it. This was the case for several young men in the small village of Bottrighe who, in early 1809, contracted civil and not religious marriages in order to avoid conscription. When, at the end of the year, local officials drew up the draft lists, these young men could not prove that they were married: the local civil registers had been burned during the year's anti-tax revolts. They were thus entered into the lists as bachelors, and had to go through an appeals process to obtain the privilege they had hoped to obtain simply by contracting civil marriage.[52]

This registration practice, and the fact that it developed at the same time in different parts of the kingdom, is quite revelatory of the conception Italians had of civil marriage – and of the registration thereof. In the popular mindset, religious marriage was still the only binding source of obligations for bride and groom. It was no accident that parties, receptions, and other forms of celebration were associated with the religious, and not the civil, ceremony.[53] Even if, in most cases, the civil and religious ceremonies took place on the same day (or nearly), contracting civil marriage could also sometimes be a useful strategy to avoid recruitment. Apparently, the authorities' persistent propaganda about the benefits of civil registration had not influenced popular mentality enough. Popular cultures of identification gave primacy to religious marriages, which people saw as the only ceremony that really mattered. Civil marriage, on the contrary, was seen as a mere formality that could also be used to deceive the state itself.

This approach to civil marriage is confirmed by another registration practice that developed in the valleys that bordered the Swiss Confederation. Here, several couples held religious marriage ceremonies on the other side of the border, in order to avoid civil registration. When he discovered that this was happening in the Valchiavenna, the prefect of the department of the Adda explained that this practice was aimed at bypassing impediments set by the Italian laws on marriage. The couples who crossed the border to marry would have not been able to contract civil marriage in the Kingdom of Italy, either because of the kinship ties between them, because of their age, or because their families had not given them permission. They thus got parish priests of the Swiss canton of Ticino to marry them, and then came back to their towns and villages as married couples.[54] Much to the chagrin of local officials, the couples' communities saw their religious marriages, celebrated abroad, as fully legitimate.[55] The Italian authorities thus urged their Swiss counterparts to forbid the practice, while asking local bishops to corroborate their efforts.[56] However, when the government effectively made it impossible for the Valchiavenna's inhabitants to marry in Ticino, the latter started to do so in the other Swiss canton that bordered the valley, Grisons.[57] Far from being "ignorant", people exploited the dual nature of the civil-religious marriage system entailed by Napoleonic legislation to bypass both religious and civil impediments. This was possible because people tended to consider religious marriage as the only real form of marriage. In the eyes of these couples, a marriage celebrated abroad, even if it was not noted in the civil registers, was an effective – and complete – form of marriage.

The separation between civil and religious forms of marking and registering important life events was also taken advantage of in another way. In February 1811, the prosecutor-general of Chioggia received a report that was destined to make its way to the minister of the Interior himself. The report – probably written by a municipal official in Chioggia – explained that some infants were being given one name on the civil register, but then being baptised with another one. The name given at baptism was usually the one that was used at home and in public. Moreover, there were also discrepancies in the surnames and domiciles registered for these children. The author of the report had ordered a comparison of the town's civil and parish registers, finding that, for 1810 alone, there were at least a hundred discrepancies between acts noted in the two registers. He had then sent the list of the discrepancies to the prosecutor, suggesting that the differences had to be corrected.[58]

When he was informed of the case, the prefect of the Adriatico complained that there was little that could be done to rectify 'such important variations'. He lamented the way that parents' 'ignorance and inaccuracy' could potentially harm their children's interests. However, as in other cases, the prefect did not call for an investigation that could uncover the reasons underlying the behaviour that drove this practice.[59] When the case reached the central government, it once again reacted by resorting to the clergy. The minister of Worship sent a circular to the bishops stating that parish priests 'must religiously observe that children be given the same name both on the civil register and at baptism'.[60] Priests, however, usually baptised newborns before they were taken for civil registration. They were therefore of little use in helping to tackle this registration practice. In this case, state authorities were ultimately powerless in the face of a specific use of the registration system that they had not foreseen.

It was religious registration that provided real social value for the ordinary people of Chioggia. The names they gave their children at baptism were the only ones they really used; "civil" names were seen as mere formalities. The number of discrepancies – over 100 out of a total of approximately 400 registrations of male newborns in 1810[61] – shows a widespread social practice the ultimate reasons of which unfortunately remain obscure. Once again, officials reductively ascribed this behaviour simply to people's "ignorance", depriving us of the sources needed to understand why they were acting this way. However, it may be possible to assume that this registration practice was another form of draft resistance. As mentioned above, Chioggia was one of the epicentres of the practice of contracting civil marriage to avoid conscription. In this sense, it might be that the practice of registering newborns with names that were not the ones they were christened with was a way to make it harder for them to be drafted in the future. If this were true, what the civil authorities had uncovered was a practice motivated by a long-term agenda.[62] In Chioggia, people were thinking decades into the future when cheating the state with its own tools.

Even if this case from Chioggia eventually resulted in a circular being sent to all the kingdom's bishops, this particular form of registration practice was probably quite rare. Of the 330 births registered in Cartura between 1806 and 1811, there are only two cases in which completely different names were given for the civil and religious registers. The small discrepancy can be explained by the significant age of the local parish priest, who may have made mistakes when transcribing names. Similarly, in Enego, of the 428 births registered between 1809 and 1813, only one is of an infant whose name is different in the two registers. Of course, this does not mean that Chioggia was an exception, and that this registration practice was wholly absent in some of the kingdom's other towns and villages. However, it was not detected beyond Chioggia, and probably remained a rare phenomenon in other contexts, where it never troubled the civil authorities.

Slight variations between the names given for civil and religious registration (such as Angelo/Arcangelo or Domenico/Giandomenico) were much more common. Evidently, in these cases, it is unlikely that the parents' ultimate aim was to save their children from being drafted in the future. These kinds of discrepancies resulted from behaviours and agencies that are very hard to explain – in some cases, they could have been simply the result of mistakes made by civil and religious registrars when transcribing these names. However, in some cases, it is possible to glimpse the rationale behind these variations. This, for instance, is the case for a newborn whose birth was registered in November 1810 in Enego. The infant's name was given as 'Angelo Napoleone' for his christening, and as 'Napoleone Angelo' at the civil registry office. This is an early case of a political name choice, and features an explicit differentiation between the names chosen for the two registration systems. At the civil registry, the element of Napoleonic tribute in the child's name was prioritised, while his religious name earned pride of place at his baptism.[63] Thus, people could use the two-step process of registration in even more unexpected ways.

*

In May 1812, the Italian minister of Finance received a worrying report from an official based in Fabriano, in the department of the Musone, according to which many local inhabitants did not have a surname. In the following months, a departmental inquiry confirmed the report's findings. More than 1,000 people were effectively devoid of a surname. One year later, following the example set by the French Empire, the Italian government issued a decree that required every citizen to take a surname. According to the Milanese authorities, this widespread lack of surnames was a legacy of the 'barbaric domination of the Lombards' – who ruled the area after the fall of the Roman Empire.

As Roberto Bizzocchi has shown in a detailed reconstruction of the case, what the state authorities perceived as an absence of surnames was actually a lack of stability in the surnames people used. In fact, the same people or families used different surnames for different forms of registration. When the government's decree forced people to choose one single surname, they reacted in various different ways. Some choose their patronymic as their surname, while others preferred to use names related to maternal kinship. According to Bizzocchi, others relished the occasion, 'with the special pleasure produced by the rare experience of choosing some part of their own destiny' This, for example, was the case of a woman named Bibiana from Cancelli, a small village located in a valley a few kilometres from the larger centre of Fabiano.

When prompted to choose a surname for herself, Bibiana opted for the rather bizarre choice of 'Cuccodrilli', the Italian word for crocodiles.[64]

This case, just like the others mentioned in this chapter, is not just a confirmation of the originality of the strategies and choices that Italians adopted in the face of the state's increasing presence in their everyday lives. It also shows how state authorities and the general population drew on different cultures of identification when it came to defining the identity of individuals. What the authorities perceived as "ignorance" was actually a manifestation of the gulf that separated these two cultures of identification. The culture of the Italian population was one in which religious ceremonies and forms of registration provided a sufficient legal basis for the development of their social lives. They thus reacted with indifference to procedures that the authorities, on the other hand, saw as essential. From the point of view of ordinary people, the new forms of registration and the duties attached to them were largely useless, when not burdensome for them. Italians thus only adopted the new procedures when it became necessary to do so in order to be able to celebrate religious ceremonies. Open resistance was not worth much of an effort, so long as the new duties remained compatible with their own culture of identification. However, people continued to avoid these new duties when the state's weakness or lack of interest made this possible. Furthermore, they also found creative ways to use the new procedures to frustrate some of the state authorities' other aims, as was the case for the civil marriages celebrated to avoid conscription.

This cleavage between two separate cultures of identification, and their failure to appreciate it, led the authorities to overestimate the scale of registration avoidance. Throughout the Napoleonic period, they continued to receive reports related to registration practices that had nothing to do with the original aims of the new legislation. Officials tended to embed these cases in a pessimistic narrative, according to which people's persistent "ignorance" inevitably meant that the state's efforts at rationalisation were unappreciated and misunderstood. This led them to adopt a stance that was dismissive of the possibilities of changing this state of affairs. Little could be done to "explain" the importance of the new legislation to people. Even efforts to mobilise the help of parish priests had not been enough to enlighten the minds of ordinary Italians. Repression would have provoked discontent, but would not have changed people's mentalities. According to officials, only time would lead Italians to fully embrace and appreciate civil registration.

In one thing, the Italian authorities were probably right: ordinary people could hardly understand the usefulness of civil registration. But this was not – as the authorities believed – the result of "ignorance" or a lack of rationality. The common people simply recognised that the new registration duties were part of a political agenda that was not their own. Their active indifference was a natural result of this mutual incomprehension.

Notes

1 Copy of a letter from the justice of the peace of Conselve to the prosecutor-general of the Brenta (4 May 1809), ASMi, Popolazione p.m., b. 6.
2 Letter from the prefect of the Tronto to the justice minister (18 July 1809, n. 3/11670), *ibid.*
3 Letter from the prosecutor-general of Padua to the justice minister (12 September 1809, n. 2146), *ibid.*
4 Michael Broers, "The Parochial Revolution: 1799 and the Counter-Revolution in Italy," *Renaissance and Modern Studies* 33 (1989): 159–173; Broers, *The Napoleonic Empire in Italy*, 58–71. For an historiographical overview: Rao, "Napoleonic Italy," in particular 86–89.
5 Report from the interior minister to the viceroy (18 May 1811, n. II/11088), ASMi, Popolazione p.m., b. 7.
6 *Codice Civile di Napoleone*, 17. The Italian regulations pointed out that 'in this case it will be necessary for the newborn to be presented to the registrar', stressing the mandatory nature of this provision: *Bollettino*, 1806a, 168–223.
7 Circular from the interior minister (5 December 1806, n. 14344), in *Raccolta delle istruzioni diramate*, 26–28.
8 Circular from the interior minister (7 December 1807, n. 18331), in *ibid.*, 25.
9 Letter from the secretary general of the Interior ministry to the minister of Worship (5 December 1806, n. 14344/II), ASMi, Popolazione p.m., b. 4.
10 Letter from the interior minister to the minister of Worship (21 December 1807, n. 29620/II), ASMi, Popolazione p.m., b. 5.
11 Circular from the interior minister to the prefects (14 April 1809, n. 8277), ASMi, Popolazione p.m., b. 6.
12 Letter from the prefect of the Tagliamento to the interior minister (20 August 1810, n. 14207).
13 Letter from the prosecutor-general of Brescia to the minister of Justice (30 August 1809, n. 5973), ASMi, Popolazione p.m., b. 6.
14 Letter from the prefect of the Adriatico to the interior minister (21 November 1808, n. II/21007), ASMi, Popolazione p.m., b. 182.
15 Letter from the vice-prefect of Campo San Piero to the interior minister (28 November 1807, n. 2867), ASMi, Popolazione p.m., b. 74.
16 Letter from the judge of first instance of Venice to the prefect of the Adriatico (21 January 1807, n. 24879), ASVe, Prefettura, b. 90.
17 Draft letter from the prefect of the Adriatico to the *savj* municipal Council (27 February 1807); letter from the *savj* municipal Council to the prefect of the Adriatico (19 February 1807, n. 752), both in *ibid.*
18 *Codice dei delitti e delle pene*, 115.
19 Letter from the justice minister to the minister of Worship (12 December 1810, n. 25898/II), ASMi, Popolazione p.m., b. 6.
20 Circular sent by the interior minister to the prefects (19 December 1810, n. II/27244), ASVe, Prefettura, b. 286.
21 Sentence (15 March 1814, n. 553), ASVi, Tribunale civile e correzionale di prima istanza, Protocollo Corte Penale 1814. See also the sentence against Francesco Aldighieri (15 March 1814, n. 461), *ibid.*
22 'Regolamento per l'attivazione in tutto lo Stato dei Registri delle nascite, de' matrimonj e delle morti,' *Bollettino*, 1806a, 168–223.
23 Draft letter from the interior minister to the prefect of the Passariano (18 February 1810, n. I/1727), ASMi, Popolazione p.m., b. 80.
24 Circular from the prefect of the Adriatico to the parish priests and mayors (6 May 1809, n. II/7453), ASMi, Popolazione p.m., b. 73.

25 He also asked the prefect not to change the wording, as this would have attracted even more attention on the matter: draft of a confidential letter from the interior minister to the prefect of the Bacchiglione (20 May 1809, n. II/11884), *ibid.*
26 Tassoni, *Arti e tradizioni popolari*, 309–310.
27 See *ibid.*
28 Letter from the prefect of the Tagliamento to the interior minister (20 August 1810, n. 14207), ASMi, Popolazione p.m., b. 85.
29 Letter from the prefect of the Adriatico to the interior minister (20 February 1811, n. 3299), ASMi, Popolazione p.m., b. 70. While the comparison operations were ongoing, the prefect also asked the departmental health committee whether any medical explanations could justify this figure. The doctor appointed by the committee asked for several elements of statistical data and concluded that there were no medical factors at play in the incongruity: letter from the departmental health committee to the prefect of the Adriatico (17 April 1811), ASVe, Prefettura, b. 354.
30 Unsigned and undated annotation, *ibid.*
31 A similar cleavage was also observed by authorities in the French Empire: Noiriel, "L'identification des citoyens," 10–11.
32 See, for instance, the circular from the archbishop of Udine to the parish priests (20 January 1809), ASMi, Popolazione p.m., b. 5.
33 The Enego civil registers are stored in ASVi, Stato civile napoleonico, Enego, 336.2, 336.3, 336.4, 336.5, 336.6. The religious ones are stored in the local parish archives. The Cartura civil registers are in in ASPa, Stato civile napoleonico, Cartura, b.es 581, 583, 584. The religious ones are in the Cartura parish archives.
34 Letter from the registrar of Torcello to the prefect of the Adriatico (24 November 1806, n. 16), ASVe, Prefettura, b. 90.
35 Letter from the mayor of Fossalta di Piave to the vice-prefect of Portogruaro (10 June 1812, n. 671), both in ASVe, San Donà, b. 56.
36 Letter from the prefect of the Tagliamento to the prefect of the Adriatico (25 August 1806, n. 628), ASVe, Prefettura, b. 26.
37 Copy of a letter from the prosecutor of Schio to the prosecutor-general of Vicenza (17 July 1812, n. 1699), ASMi, Popolazione p.m., b. 8.
38 Letter from the prosecutor-general of Vicenza to the minister of Justice (22 July 1812, n. 3410), *ibid.*
39 Draft letter from the minister of Justice to the minister of Worship (29 July 1812, n. 16469), *ibid.*
40 Letter from the bishop of Concordia to the prefect of the Adriatico (23 June 1811, n. 182), ASVe, Prefettura, b. 383; letter from the bishop of Concordia to the minister of Worship (8 July 1811, n. 191), ASMi, Popolazione p.m., b. 54.
41 See also Andrea Schiaffino, "Coscrizione e nuzialità in età napoleonica," in *Reggio e i territori estensi dall'antico regime all'età napoleonica*, eds. Marino Berengo and Sergio Romagnoli (Parma: Pratiche, 1979), vol. 1, 179–196.
42 Conscription law (13 August 1802), *Bollettino*, 1802, 234–252.
43 Letter from the minister of war to the interior minister (26 January 1809, n. 22). The proposal was opposed by the Justice ministry, who considered it 'harsh and vexatious': draft report from the II division of the Justice ministry (30 February 1809), both in ASMi, Popolazione p.m., b. 54.
44 Letter from the mayor of Chioggia to the vice-prefect of Chioggia (31 March 1809, n. 521), ASVe, Prefettura, b. 206.
45 Letter from the prosecutor-general of Como to the justice minister (29 August 1809, n. 1442), ASMi, Popolazione p.m., b. 54.
46 Letter from the justice of the peace of Carpi to the prosecutor-general of Modena (6 March 1810, n. 64), ASMi, Popolazione p.m., b. 54.

47 The Council proposed applying the provision only to those who clearly had married older women to avoid conscription: session of the Council of State (24 July 1810), ASMi, Registri, vol. 633, 107.
48 Circular from the minister of Worship to the bishops (8 October 1810, n. I/11896), ASMi, Popolazione p.m., b. 54.
49 Letter from the prefect of the Adige to the justice minister (9 December 1810, n. III/33738), *ibid.*
50 Civil register of marriages, ASPa, Stato civile napoleonico, Cartura, 583; parish register of marriages, Cartura parish archives.
51 Unsigned and undated annotation, ASVe, Prefettura, b. 354.
52 Letter from the mayor of Bottrighe to the vice-prefect of Adria (6 November 1809, n. 257); letter from the mayor of Bottrighe to the vice-prefect of Adria (6 November 1809, n. 258), both in ASMi, Popolazione p.m., b. 70.
53 See the dozens of reports from parish priests on wedding practices in Tassoni, *Arti e tradizioni popolari.*
54 Letter from the prefect of the Adda to the interior minister (7 September 1811, n. III/10406), ASMi, Popolazione p.m., b. 55.
55 Letter from the interior minister to the justice minister (13 September 1811, n. II/21680), *ibid.*
56 Decree of the Small Council of the Canton Ticino (23/110/1811); letter from the vicar general of the diocese of Milan to the minister of Worship (30 September 1811, n. 242), both in *ibid.*
57 Letter from interior minister to the justice minister (11 July 1812, n. II/19234), *ibid.*
58 Report quoted in the letter from the vice-prefect of Chioggia to the general police commissioner of Venice (2 February 1811, n. 473), ASVe, Prefettura, b. 354.
59 Draft letter from the prefect of the Adriatico to the interior minister (27 February 1811, n. 11), ASVe, Prefettura, b. 354.
60 Circular from the minister of Worship to the bishops (29 March 1811, n. 19707/I), ASMi, Popolazione p.m., b. 182.
61 The Chioggia civil registers contain a total of 867 birth records for 1810: Chioggia registers of births (1810), ASVe, Stato civile napoleonico, 476.1 and 476.2.
62 Similarly, in circular sent in 1809, the bishop of Belluno suggested that the ultimate cause of the lack of registration of births was to 'save [the newborns] from conscription, if their names cannot be found registered in the acts of the civil registration': circular from the bishop of Belluno to the parish priests (15 January 1809), ASMi, Popolazione p.m., b. 5. Even if this aim cannot be excluded, no other hints of it can be found in the administrative sources consulted for this book. Also in this case, it is apparent that the authorities' "selective blindness" towards people' behavior makes it much harder for historians to understand registration practices.
63 Curiously enough, it was the village chaplain himself – who had christened him as Angelo Napoleone – who registered the newborn at the civil registry as Napoleone Angelo: ASVi, Stato civile napoleonico, b. 336.3. On the political naming of newborns in modern Italy, see Stefano Pivato, *Il nome e la storia. Onomastica e religioni politiche nell'Italia contemporanea* (Bologna: il Mulino, 1999).
64 Bizzocchi, "Marchigiani senza cognome."

Conclusion

The end of Napoleonic rule in 1814 implied both changes and continuity in the state procedures related to citizens' personal identities. In Habsburg Lombardy-Venetia, the new government largely built on the administrative tradition of the Kingdom of Italy, facing similar difficulties and political issues.[1] Even when identification procedures changed, however, the new authorities continued to be inspired by a similar culture of identification.

Following the extension of the imperial marriage law to the Italian territories, civil marriage was abolished, with the religious ceremony taking on civil validity too. Parish registers of marriage thus became the only legal documents recognised by the state in this domain.[2] In December 1815, the Habsburg governor took charge of the civil registers of births and deaths. From that moment on, these registers were also entrusted to the parish priests – who managed them alongside the religious registers of baptisms and burials.[3] This solution was clearly a clean break with the Napoleonic conception of a civil registration system managed directly by state agents. However, in practical terms, it implied a continuation of the lower clergy's previous cooperation in the management of the civil registers. The political climate of the Restoration, moreover, did not diminish the state's will to control the registration of its subjects' identities. What had started as a pragmatic, on-and-off form of cooperation between the state and the church in the management of people's identities now evolved into a system of structural and political cooperation based on compatible cultures of identification.

This structural continuity was also reflected in the substantial stability of the regulations concerning identity documents. The Habsburg authorities continued to exert control over people's identity and mobility via a dual system based on registration certificates and internal passports. The system established by the Italian reform of 1811 was thus maintained in the following decades, confirming the close relationship between the registration of domicile and the identification of individuals. Moreover, the administrative limitations to suspects' movements embodied by the *module* continued to exist well after the Restoration.[4] The Habsburg authorities continued to struggle with the difficult balancing act entailed by their will to both control people's mobility and guarantee convenience for travellers. However, they did

DOI: 10.4324/9781003412434-8

not begin to liberalise the system until 1850. In that year, internal passports were abolished for travel within Venetia. This measure had been made inevitable by the opening of the railway line between Marghera and Padua a few years earlier.[5] The introduction of railways profoundly changed mobility patterns and made the system which the Napoleonic authorities had elaborated four decades earlier both politically and practically impossible to maintain.

The administrative changes that took place under French influence did indeed have long-term consequences for the ways in which different states controlled Italian societies. In this sense, this book confirms the important role the two decades of French rule played in the shifting power relations in the Italian peninsula. This was a period of profound innovation in the ways in which states – and the élites that controlled them – related to the rest of society. These changes were based on a conception of state authority that had deep roots in the enlightened absolutism promoted also by eighteenth-century Italian states. The influence of the French experience, however, was crucial in accelerating these changes. The pace of administrative reforms sped up during the 20 years in which the Italian peninsula was under the influence of post-revolutionary France. The history of personal identification, as we have seen throughout this book, confirms this.

However, this book also sheds light on aspects of this acceleration that tend to be overlooked by the 'modernisation paradigm' that has dominated the literature on this period of Italian history. Firstly, the role of French influence seems less important if we shift our point of view from the *départements réunis* to the Cisalpine Republic and the Kingdom of Italy. Here, administrative reforms were conceived of and implemented by political élites who had been deeply influenced by the native Habsburg administrative tradition. Especially in the first decade of French influence in Northern Italy (c. 1796–1805), this legacy remained important in shaping the decisions and ideas of Cisalpine and Italian politicians. After the monarchic turn, French influence became more important, especially with the adoption of the imperial Codes. However, French and Habsburg sources of administrative knowledge coexisted throughout the whole period. This duality was fully embodied in a political apparatus that was almost entirely composed of 'native' Italians. As supporters of post-revolutionary reformist stances, these politicians were both willing to adopt the novelties coming from France and eager to adapt these to the Italian context. The autonomy of the Cisalpine and Italian states with respect to France was certainly limited, but the local élites tried to make as much use of it as possible. The result was an administrative experience that cannot simply be labelled as stemming directly from French domination.

Moreover, the changes in personal identification – and more broadly in the relations between state and society – cannot be seen as emanating solely from the will and actions of the ruling classes. The Italian case study shows that, by adopting multiple points of view, it is possible to recover the agency and the importance of other historical actors in the processes that affected European societies during the post-revolutionary and Napoleonic period.

Bureaucrats were crucial in establishing how the state identification culture was actually implemented on the ground. Napoleonic administrations are commonly seen as hierarchical organisations in which obscure bureaucrats answered efficiently to orders coming from authorities higher up in the chain of command. The reality, however, was much more complex. State agents – officials, registrars, and policemen, among others – actively interpreted the orders they received from above, adapting them according to their understanding and convenience. Authorities at the higher levels, in turn, had to come to terms with the autonomy displayed by their subordinates. They had neither the economic resources nor the technological knowledge necessary to effectively control how their orders were implemented on the ground. The state's weakness allowed its agents a large degree of discretion. The latter, in turn, used their discretional powers to adapt and change the civil identification and registration systems to fit their own mentalities and work to their advantage.

Moreover, to make the system work, the state had to resort to individuals who were beyond its jurisdiction. Especially in the case of civil registration, this necessity was confirmed by the appointment of registrars and adjunct registrars who were not municipal officials. Even if these individuals were theoretically now part of the administrative system, in practice, they continued to think and act primarily as private citizens, often putting their interests ahead of those of the state. Nonetheless, without the help of these part-time bureaucrats, the entire political project of civil registration would have been simply unsustainable.

Many of these individuals were clergymen. Michael Broers has notably described what happened in Italy during the Napoleonic period as a 'war against god', highlighting how the modernisation project promoted by the French and the élites that supported them came into conflict with Catholic culture and the church.[6] Yet, once again, if we shift our perspective from the *départements réunis* to the Kingdom of Italy, we find something quite different. Church personnel was heavily involved in the personal identification procedures created by the state. Parish priests often explained and justified these new procedures from the pulpit. When needed, they acted as registrars and *aggiunti* for civil registration procedures. Their *fedi* became the basis for the new state identity documents: without them, it was impossible to obtain a security card or a passport. Moreover, the Italian bishops provided substantial ideological support to the new laws by developing political and theological arguments in their favour. Even when Napoleon broke with and imprisoned the pope, clergymen continued their precious and discreet cooperation at the local level. This was an impressive contribution for a church that was supposedly in open conflict with the state.

Last but not least, ordinary people played a crucial role in shaping how identification and registration worked. The Italian population quickly understood the limits of the innovations promoted by the state. From the beginning, it became clear that the authorities did not have the resources to make the new rules enforceable at all times and in all places. Moreover, unlike mass conscription and new taxes, the new personal identification measures were not too intrusive in the everyday life of people and their communities. This is why the use of security cards was quickly abandoned in the Italian countryside. It also explains why parents did not always rush to register the births of their children in the municipal registers. Despite the authorities' considerable efforts to explain the usefulness of these new measures, they did not manage to get ordinary people to comply with their political projects. The new regulations on documents and registration were experienced with a sense of detachment. Their alleged benefit remained intangible for the common people. Officials interpreted this attitude as a sign of 'ignorance', even though it was actually a demonstration of indifference to the Napoleonic authorities' political projects.

But indifference did not mean passivity. In fact, people elaborated a variety of strategies and behaviours to cope with the new requests coming from the state. They adopted a situational attitude to identity documents, ignoring them when they were unlikely to be stopped by the police. They registered their vital acts in the municipal registers, but when this was not strictly necessary (as for births), they did not always comply with the rules. At the same time, ordinary people continued to obtain proof of identity and to register their vital acts as they had always done in the past. Parish certificates and registers remained a major source of identity in Napoleonic Italy. People continued to see these traditional means of identification as a more relevant – and probably more reliable – way of certifying their identities. This 'active indifference' was the product of the persistence of a communal identification culture that survived the reforms promoted by state authorities.

These latter, in turn, never tried to completely supplant the traditional procedures of identification. On the contrary, the two systems coexisted throughout the Napoleonic period. The state created a parallel system of registration to the parish one. Similarly, it did not try to replace the parish certificates with its own identity documents. The establishment of a state system of identification and registration was not seen as totally incompatible with traditional identification procedures. The strategy adopted by the state authorities was more gradual. It reflected the élites' underlying recognition that it would have been impossible to completely replace the traditional system in the short term. If the civil authorities had eliminated the older means of identification and registration, they would not have been able to quickly provide a valid and effective alternative. Thus, the new system and procedures operated side by side with the old ones, without abolishing them.

This gradualism, however, did not stem from a lack of boldness or ambition on behalf of the state. Its ultimate aim was indeed to replace the traditional system of identification and registration with new one, managed and controlled by the civil authorities. But this new system had to be implemented on the ground first. The state thus devoted its energy to this demanding task. At the same time, however, the civil authorities began to assert the legal value of their own system of registration and identification. In this way, they created a theoretical hierarchy between their system and the traditional one. The latter was not eliminated, but its legal value was called into question. In this case too, the civil authorities' behaviour was inspired more by a wish to achieve hegemony than by a desire to trigger an open conflict with older habits. The authorities pragmatically accepted a pluralism in the means of identification and registration, but they immediately claimed that their own system was the only valid one in legal terms.

This battle for social hegemony was far from being resolved at the end of the Napoleonic era. In the following decades, while the legal culture of identification became stronger, as states developed their own structures for establishing and certifying identities, the communal culture of identification continued to be bolstered by a strong sense of inertia among a large part of the population. However, the forms of identity promoted by the authorities slowly came to prominence. The type of personal identity that the state certified with registers and documents started to be seen as inevitable and, somehow, natural. Individuals partially lost (or renounced) the ability to change and adapt their identities according to their convenience and interest. By the twentieth century, the state's jurisdiction over personal identity had become predominant throughout European societies.[7]

However, this hegemony did not imply that other ways of certifying one's identity disappeared. In fact, the state could only enforce the *legal* value of the identities it created, but not their universal *social* value. In 1926, a man in his forties was brought to a mental asylum in Collegno, not far from Turin. He claimed to be incapable of remembering who he was. As a consequence of the case's media coverage, he was recognised by his alleged wife as Giulio Cannella, a professor and philosopher who had disappeared during the First World War. However, the authorities were not convinced that this was his identity. After several legal proceedings that involved the close examination of fingerprints and photographs, they declared that he was actually Mario Bruneri, a criminal who had been arrested several times over the previous years.[8] Despite the authorities' decision, Cannella/Bruneri's wife continued to consider the man to be her long-lost husband after he was released from prison in 1933. They moved to Brazil with their children and lived together until 'professor Giulio Cannella' died in 1941. Several decades later, in 1970, the Vatican secretary of state issued a document confirming that the man was indeed Giulio Cannella – and that his marriage and offspring were therefore legitimate.[9]

The identity that the state made great efforts to certify for Cannella/Bruneri was not recognised by his family or by the church. Establishing a complete monopoly over the means of identification proved to be an unachievable political project even for a state as authoritarian as fascist Italy. As this and other cases show, studying changes in the history of personal identification in modern Europe does not simply result in a story charting the inevitable rise of the state. On the contrary, it allows us to explore the ever-dynamic fault lines in the relationship between states and societies. This "conflict of identities" is still alive in our current and increasingly digital lives.

Notes

1 Marco Meriggi, *Il Regno Lombardo-Veneto* (Torino: Utet, 1987), 33–105.
2 Decrees by the president of the Government Regency (1 July 1815), in *Atti del governo. Parte prima. Dal 1° gennajo al 30 settembre 1815* (Milano: Dalla C.R. Stamperia di Governo, s.d.), 99–129.
3 Decree and regulations by the Lombardy-Venetia governor (19–20 December 1815, n. 29667/1341), in *ibid.*, 1112–1120.
4 Fabio Bortoluzzi, "Il precetto politico nelle province venete (1813–1850)," in *Amministrazione della giustizia penale e controllo sociale nel Regno Lombardo-Veneto*, eds. Giovanni Chiodi and Claudio Povolo (Sommacampagna: Cierre, 2007), 271–289.
5 Andrea Geselle, "Domenica Saba Takes to the Road: Origins and Development of a Modern Passport System in Lombardy-Veneto," in *Documenting Individual Identity: The Development of State Practices in the Modern World*, eds. Jane Caplan and John Torpey (Princeton, NJ-Oxford: Princeton University Press 2001), 199–217.
6 See Broers, *The Politics of Religion in Napoleonic Italy*.
7 See for instance Higgs, *Identifying the English*, 122–186; Noiriel, "L'identification des citoyens," 27; John Torpey, "The Rise of States and the Regulation of Movement," in *Procedure, metodi, strumenti per l'identificazione delle persone e per il controllo del territorio*, ed. Livio Antonielli (Soveria Mannelli: Rubbettino, 2014), 185–196.
8 Lisa Roscioni, *Lo smemorato di Collegno. Storia italiana di un'identità contesa* (Torino: Einaudi, 2007).
9 Natalia Aspesi, "L'eterno smemorato," «La Repubblica», 21 October 1988.

Bibliography

Primary Sources

Archivio di Stato di Bologna

Archivio napoleonico, boxes: IX/27; IX/28; IX/29; IX/30; X/6; XI/26; XI/124; XI/303.

Ufficio di polizia del Reno, boxes: 32; 35; 84; 118.

Prefettura del dipartimento del Reno, boxes: 1803, tit. XIV, rub. 3; 1804, tit. XX, rub. 10–17; 1805, tit. XX, rub. 10–19; 1806, tit. XX, rub. 10–11; 1807, tit. XX, rub. 10; 1807, tit. XX, rub. 11; 1808, tit. XX, rub. 1–5; 1808, tit. XX, rub. 10; 1808, tit. XX, rub. 11; 1809, tit. XX, rub. 10; 1809, tit. XX, rub. 11, I; 1809, tit. XX, rub. 11, II; 1810, tit. XX, rub. 10, I; 1810, tit. XX, rub. 10, II; 1810, tit. XX, rub. 11, I; 1810, tit. XX, rub. 11, II; 1811, tit. XX, rub. 10, I; 1811, tit. XX, rub. 10, II; 1811, tit. XX, rub. 11; 1812, tit. XX, rub. 11–20; tit. XX, rub. 10, I; 1812, tit. XX, rub. 10, II; 1812, tit. XX, rub. 11; 1813, tit. XX, rub. 10; 1813, tit. XX, rub. 11.

Archivio di Stato di Milano

Atti di Governo, Culto parte antica, boxes: 2173; 2222.

Atti di Governo, Culto parte antica, boxes: 86.

Atti di Governo, Popolazione parte antica, box: 16.

Atti di Governo, Popolazione parte moderna, boxes: 1; 2; 3; 4; 5; 6; 7; 8; 9; 10; 51; 52; 53; 54; 55; 69; 70; 73; 74; 80; 85; 101; 182; 183.

Miscellanea storica, box: 53.

Registri di verbali e atti di governo e deliberazioni, volumes: 596; 597; 602; 603; 604; 606; 608; 609; 610; 611; 613; 617; 619; 621; 627; 633; 637; 638; 641b; 642.

Archivio di Stato di Venezia

Prefettura del dipartimento dell'Adriatico, boxes: 26; 90; 206; 207; 226; 284; 285; 286; 304; 354; 382; 383; 461; 563.

Viceprefettura di Chioggia, boxes: 34; 39.

Viceprefettura di San Donà (poi Portogruaro), boxes: 8; 56; 80; 81.

Archivio di Stato di Vicenza

Commissariato di polizia, boxes: 1; 2; 3; 4; 5; 6; 7; 8; 9; 10; 11; 12; 13; 14; 15; 16; 17; 18; 19; 20; 21; 22; 23; 24; 25; 26; 27; 28; 29.
Corte di giustizia civile e criminale, boxes: 2; 4; 12; 14.
Protocollo Corte Penale, volumes: 1807–1809; 1810–1811; 1814.
Protocollo Tribunale Correzionale, volumes: 1807–1809; 1810–1811.

Biblioteca Civica Bertoliana

Archivio storico del Comune di Vicenza (archivio Torre), volumes: 992; 993; 994; 995; 2109; A9; A10; A11; A12; A13; A14; A15; A16; A17.
Arnaldo I Arnaldi Tornieri, Memorie di Vicenza che cominciano dall'anno 1767, 18 giugno e terminano al 1822, Biblioteca Civica Bertoliana, ms. 3108–3111.

Published Parliamentary Sources

Il redattore del Gran Consiglio della Repubblica Cisalpina, n. 1–358 (1797–1798).
Processo verbale delle sessioni del Consiglio de' Seniori, n. 1–212 (1798–1799).
Processo verbale del Consiglio de' Juniori, n. 1–114 (1798–1799).

Assemblee

Assemblee della Repubblica cisalpina, vol. 1, parte 1, edited by Camillo Montalcini and Annibale Alberti (Bologna: Zanichelli, 1917).
Assemblee della Repubblica cisalpina, vol. 1, parte 2, edited by Camillo Montalcini and Annibale Alberti (Bologna: Zanichelli, 1917).
Assemblee della Repubblica cisalpina, vol. 2, edited by Camillo Montalcini and Annibale Alberti (Bologna: Zanichelli, 1917).
Assemblee della Repubblica cisalpina, vol. 3, edited by Camillo Montalcini and Annibale Alberti (Bologna: Zanichelli, 1919).
Assemblee della Repubblica cisalpina, vol. 4, edited by Camillo Montalcini and Annibale Alberti (Bologna: Zanichelli, 1919).
Assemblee della Repubblica cisalpina, vol. 5, edited by Camillo Montalcini and Annibale Alberti (Bologna: Zanichelli, 1927).
Assemblee della Repubblica cisalpina, vol. 6, edited by Camillo Montalcini and Annibale Alberti (Bologna: Zanichelli, 1927).
Assemblee della Repubblica cisalpina, vol. 7, edited by Annibale Alberti, Roberto Cessi, and Lorenzo Marcucci (Bologna: Zanichelli, 1935).
Assemblee della Repubblica cisalpina, vol. 8, edited by Annibale Alberti, Roberto Cessi, and Lorenzo Marcucci (Bologna: Zanichelli, 1938).
Assemblee della Repubblica cisalpina, vol. 9, edited by Annibale Alberti, Roberto Cessi, and Lorenzo Marcucci (Bologna: Zanichelli, 1940).
Assemblee della Repubblica cisalpina, vol. 10, edited by Annibale Alberti, Roberto Cessi, and Lorenzo Marcucci (Bologna: Zanichelli, 1943).
Assemblee della Repubblica cisalpina, vol. 11, edited by Annibale Alberti, Roberto Cessi, and Lorenzo Marcucci (Bologna: Zanichelli, 1948).

Published Legislative Sources

Gridario generale, ossia Collezione di tutte le gride, editti, Proclami pubblicati in Milano (Milano: [1799–1800]), 2 vol.

Codice Civile di Napoleone il grande pel Regno d'Italia. Quarta Edizione originale ed officiale. Tomo I (Milano: Dalla Reale Stamperia, 1806).

Codice di procedura civile pel Regno d'Italia (Milano: Dalla reale stamperia, 1806).

Raccolta delle decisioni, regolamenti, circolari ed istruzioni relative agli attributi proprj delle autorità amministrative nel dipartimento del Bacchiglione. Volume primo (Vicenza: Da Bartollomeo Paroni tipografo dipartimentale, 1809).

Raccolta delle istruzioni diramate dal ministero dell'interno del Regno d'Italia per l'esecuzione del decreto di S.A.I. il Principe Vicerè 27 marzo 1806 sui registri degli atti dello stato civile (Milano: Dalla Stamperia Reale, 1809).

Codice dei podestà e sindaci del Regno d'Italia (Milano: Dalla Reale Stamperia, 1811).

Raccolta delle decisioni, regolamenti circolari ed istruzioni relative agli attributi proprj delle autorità amministrative nel dipartimento del Bacchiglione. Volume quinto (Vicenza: Da Bartolommeo Paroni, 1813).

La Législation civile, commerciale et criminelle de la France ou commentaire et complément des codes francais [...] par m. le baron Locré, Tome Troisième (Paris: Treuttel et Wurtz, 1827).

Raccolta

Raccolta, 1/I: *Raccolta degli ordini ed avvisi stati pubblicati dopo il cessato Governo Austriaco* (Milano: Presso Luigi Veladini in Contrada Santa Redagonda, Anno IV).

Raccolta, 1/II: *Raccolta degli ordini, avvisi, proclami ec. Pubblicati in Milano nell'anno V Repubblicano Francese. Tomo II* (Milano: Presso Luigi Veladini in Contrada Santa Redagonda, 22 Settembre MDCCXCVI).

Raccolta, 1/III: *Raccolta degli ordini, avvisi, e proclami. Pubblicati in Milano nell'anno V Repubblicano Francese. Tomo III* (Milano: Presso Luigi Veladini in Contrada Santa Redagonda, 8 Maggio 1797).

Raccolta, 1/IV: *Raccolta delle leggi, proclami, ordini, ed avvisi. Pubblicati in Milano nell'anno VI Repubblicano. Tomo IV* (Milano: Presso Luigi Veladini in Contrada S. Redagonda, 1 Novembre 1797).

Raccolta, 1/V: *Raccolta delle leggi, proclami, ordini ed avvisi. Pubblicati in Milano nell'anno VI Repubblicano. Tomo V* (Milano: Presso Luigi Veladini in Contrada S. Redagonda, 2 Aprile V. S. 1798).

Raccolta, 1/VI: *Raccolta delle leggi, proclami, ordini ed avvisi ec. Pubblicati in Milano nell'anno VII repubblicano. Tomo VI* (Milano: Presso Luigi Veladini in Contrada S. Redagonda, 1 vendemmiale 22 Settembre 1798. V. S.).

Raccolta, 1/VII: *Raccolta delle leggi, proclami, ordini ed avvisi ec. Pubblicati in Milano nell'anno VII repubblicano. Tomo VII* (Milano: Presso Luigi Veladini in Contrada S. Redagonda, 1 Germile, 21 Marzo 1799. v. s.).

Raccolta, 2/I: *Raccolta delle leggi, proclami, ordini ed avvisi. Pubblicati in Milano dal giorno 13 Pratile Anno VIII (2 Giugno 1800) epoca del ritorno dell'Armata Francese in questa Città. Tomo I* (Milano: Presso Luigi Veladini in Contrada S. Radegonda).

Raccolta, 2/II: *Raccolta delle leggi, proclami, ordini ed avvisi. Pubblicati in Milano dal giorno 13 Pratile Anno VIII (2 Giugno 1800) epoca del ritorno dell'Armata Francese in questa Città. Tomo II* (Milano: Presso Luigi Veladini in Contrada S. Radegonda).

Raccolta, 2/III: *Raccolta delle leggi, proclami, ordini ed avvisi. Pubblicati in Milano dal giorno 12 Messidoro anno IX (1 Luglio 1801) sino a tutto il giorno 10 Nevoso anno X (31 Dicembre 1801). Tomo III* (Milano: Presso Luigi Veladini in Contrada S. Radegonda).

Leggi

Leggi, 1/I: *Leggi della Repubblica Cisalpina dal giorno dell'installamento del Corpo Legislativo. Volume primo* (Milano: Dalla stamperia italiana, e francese, a S. Zeno, N.° 534, Anno I.° della Repubblica, 1798).

Leggi, 1/II: *Leggi della Repubblica Cisalpina dal giorno dell'installamento del Corpo Legislativo. Volume secondo* (Milano: Dalla stamperia italiana, e francese, a S. Zeno, N.° 534, Anno I.° della Repubblica, 1798).

Leggi, 1/III: *Leggi della Repubblica Cisalpina dal giorno dell'installamento del Corpo Legislativo. Volume terzo* (Milano: Dalla stamperia italiana, e francese, a S. Zeno, N.° 534, Anno I.° della Repubblica, 1798).

Leggi, 1/IV: *Leggi della Repubblica Cisalpina dal giorno dell'installamento del Corpo Legislativo. Volume quarto* (Milano: Dalla stamperia italiana, e francese, a S. Zeno, N.° 534, Anno I.° della Repubblica, 1798).

Leggi, 1/V: *Leggi della Repubblica Cisalpina dal giorno dell'installamento del Corpo Legislativo. Volume quinto* (Milano: Dalla stamperia italiana, e francese, a S. Zeno, N.° 534, Anno VI.° della Repubblica, 1798).

Leggi, 1/VI: *Leggi della Repubblica Cisalpina dal giorno dell'installamento del Corpo Legislativo. Volume sesto* (Milano: Dalla stamperia italiana, e francese, a S. Zeno, N.° 534, Anno VI.° della Repubblica, 1798).

Leggi, 2/I: *Leggi della Repubblica Cisalpina dal giorno dell'attivazione della Costituzione dell'anno sesto, con i Proclami del Direttorio Esecutivo, e dei Ministri. Volume primo* (Milano: Dalla stamperia italiana, e francese a S. Zeno, N.° 534, Anno VII Repubblicano).

Leggi, 2/II: *Leggi della Repubblica Cisalpina dal giorno dell'attivazione della Costituzione dell'anno sesto, con i Proclami del Direttorio Esecutivo, e dei Ministri. Volume secondo* (Milano: Dalla stamperia italiana, e francese a S. Zeno, N.° 534, Anno VII Repubblicano).

Leggi, 2/III: *Leggi della Repubblica Cisalpina dal giorno dell'attivazione della Costituzione dell'anno sesto, con i Proclami del Direttorio Esecutivo, e dei Ministri. Volume terzo* (Milano: Dalla stamperia italiana, e francese a S. Zeno, N.° 534, Anno VII Repubblicano).

Leggi, 2/IV: *Leggi della Repubblica Cisalpina dal giorno dell'attivazione della Costituzione dell'anno sesto, con i Proclami del Direttorio Esecutivo, e dei Ministri. Volume quarto* (Milano: Dalla stamperia italiana, e francese a S. Zeno, N.° 534, Anno VII Repubblicano).

Leggi, 2/V: *Leggi della Repubblica Cisalpina dal giorno dell'attivazione della Costituzione dell'anno sesto, con i Proclami del Direttorio Esecutivo, e dei Ministri. Volume quinto* (Milano: Dalla stamperia italiana, e francese a S. Zeno, N.° 534, Anno VII Repubblicano).

Leggi, 2/VI: *Leggi della Repubblica Cisalpina dal giorno dell'attivazione della Costituzione dell'anno sesto, con i Proclami del Direttorio Esecutivo, e dei Ministri. Volume sesto* (Milano: Dalla stamperia italiana, e francese a S. Zeno, N.° 534, Anno IX Repubblicano).

Bollettino

Bollettino, 1802: *Bollettino delle leggi della Repubblica Italiana. Dalla Costituzione proclamata nei Comizj in Lione al 31 Dicembre 1802. Anno I. N. 1 al 20* (Presso Luigi Veladini Stampatore Nazionale in Contrada s. Radegonda).

Bollettino, 1803: *Bollettino delle leggi della Repubblica Italiana. Dal 1 Gennajo ai 31 Dicembre 1803. Anno II. N.° 1 al 16* (Presso Luigi Veladini Stampatore Nazionale in Contrada s. Radegonda).

Bollettino, 1804a: *Bollettino delle leggi della Repubblica Italiana. Parte prima. Dal 1 Gennajo ai 30 Aprile 1804. Anno III. N.° 1 al 11* (Presso Luigi Veladini Stampatore Nazionale in Contrada S. Radegonda).

Bollettino, 1804b: *Bollettino delle leggi della Repubblica Italiana. Parte seconda. Dal 1 Maggio al 31 Dicembre 1804. Anno III. N.° 12 al 25* (Presso Luigi Veladini Stampatore Nazionale in Contrada S. Radegonda).

Bollettino, 1805a: *Bollettino delle leggi del Regno d'Italia. Parte prima. Dal 1 Gennajo al 30 Giugno 1805. N.° 1 al 10* (Milano: Presso la Regia Stamperia Veladini in Contrada S. Radegonda).

Bollettino, 1805b: *Bollettino delle leggi del Regno d'Italia. Parte seconda. Dal 1 Luglio al 31 Dicembre 1805. N. 74, al N. 154* (Milano: Presso la Regia Stamperia Veladini in Contrada S. Radegonda).

Bollettino, 1806a: *Bollettino delle leggi del Regno d'Italia. Parte prima. Dal 1 gennajo al 30 aprile 1806. N. 1 al N. 55* (Milano: Dalla Reale Stamperia).

Bollettino, 1806b: *Bollettino delle leggi del Regno d'Italia. Parte II. Dal 1 maggio al 31 agosto 1806. Coll'aggiunta dei Decreti pubblicati negli Stati Veneti avanti la loro unione al Regno. N. 11 al N. 28* (Milano: Dalla Reale Stamperia).

Bollettino, 1806c: *Bollettino delle leggi del Regno d'Italia. Parte III. Dal 1 settembre al 31 dicembre 1806. N. 29 al N. 39* (Milano: Dalla Reale Stamperia).

Bollettino, 1807a: *Bollettino delle leggi del Regno d'Italia. Parte prima. Dal 1 gennajo al 30 giugno 1807. N. 1 al N. 16* (Milano: Dalla Reale Stamperia).

Bollettino, 1807b: *Bollettino delle leggi del Regno d'Italia. Parte seconda. Dal primo luglio al 30 settembre 1807. N. 17 al N. 28* (Milano: Dalla Reale Stamperia).

Bollettino, 1807c: *Bollettino delle leggi del Regno d'Italia. Parte terza. Dal primo ottobre al 31 dicembre 1807. N. 29 al N. 41* (Milano: Dalla Reale Stamperia).

Bollettino, 1808a: *Bollettino delle leggi del Regno d'Italia. Parte prima. Dal primo gennajo al 31 maggio 1808. N.° 1 al N.° 18* (Milano: Dalla Reale Stamperia).

Bollettino, 1808b: *Bollettino delle leggi del Regno d'Italia. Parte seconda. Dal primo giugno al 31 dicembre 1808. N.° 19 al N.° 43* (Milano: Dalla Reale Stamperia).

Bollettino, 1809: *Bollettino delle leggi del Regno d'Italia. Dal primo gennajo al 31 dicembre 1809. N.° 1 al N.° 19* (Milano: Dalla Reale Stamperia).

Bollettino, 1810a: *Bollettino delle leggi del Regno d'Italia. Parte prima. Dal primo gennajo al 30 giugno 1810. N.° 1 al N.° 10* (Milano: Dalla Reale Stamperia).

Bollettino, 1810b: *Bollettino delle leggi del Regno d'Italia. Parte II. Dal 1.° luglio al 30 settembre 1810. N.° 11 al N.° 22* (Milano: Dalla Reale Stamperia).

Bollettino, 1810c: *Bollettino delle leggi del Regno d'Italia. Parte III. Dal 1.° ottobre al 31 dicembre 1810. N.° 23 al N.° 31* (Milano: Dalla Reale Stamperia).

Bollettino, 1811a: *Bollettino delle leggi del Regno d'Italia. Parte prima. Dal 1.° gennajo al 30 giugno 1811. N.° 1 al N.° 16* (Milano: Dalla Reale Stamperia).

Bollettino, 1811b: *Bollettino delle leggi del Regno d'Italia. Parte II. Dal 1.° luglio al 31 dicembre 1811. N.° 17 al N.° 34* (Milano: Dalla Stamperia Reale).

Bollettino, 1812a: *Bollettino delle leggi del Regno d'Italia. Parte prima. Dal 1.° gennajo al 30 giugno 1812. N.° 11 al N.° 16* (Milano: Dalla Stamperia Reale).

Bollettino, 1812b: *Bollettino delle leggi del Regno d'Italia. Parte II. Dal 1.° luglio al 31 dicembre 1812. N.° 17 al N.° 22* (Milano: Dalla Stamperia Reale).

Bollettino, 1813a: *Bollettino delle leggi del Regno d'Italia. Parte prima. Dal 1.° gennajo al 30 giugno 1813. N.° 1 al N.° 15* (Milano: Dalla Stamperia Reale).

Bollettino, 1813b: *Bollettino delle leggi del Regno d'Italia. Parte seconda. Dal 1.° luglio al 31 dicembre 1813. N.° 16 al N.° 21* (Milano: Dalla Stamperia Reale).

Bollettino, 1814: *Bollettino delle leggi del Regno d'Italia. Dal 1.° gennajo al 6 aprile 1814. N.° 1 e 2* (Milano: Dalla Stamperia Reale).

Published Primary Sources

Casanova, Giacomo. *Mémoires de Jacques Casanova de Seingalt écrits par lui-même. Édition originale, la seule complète. Tome cinquième*. Paris: Ernest Flammarion éditeur, 1832.

Charvillhac. *Le guide de l'officier de l'état civil, ou Méthode simple et facile pour diriger la marche qu'il doit suivre dans la rédaction des actes de son ministère*. Paris: Chez Rondonneau, 1806.

De La Révellière-Lépeaux, Louis-Marie. *Réflexions sur le culte, sur les cérémonies civiles et sur les fétes nationales*. Paris: Chez H.J. Jansen, l'an 5me.

Giovannetti, Giacomo. *Manuale degli ufficiali dello stato civile in cui sono gradatamente tracciate le operazioni da farsi nella formazione de' singoli atti e le module per ciascun d'essi*. Novara: dalla stamperia di Giuseppe Rosario, 1809.

Locré, Jean-Guillaume. *La Législation civile, commerciale et criminelle de la France ou commentaire et complément des codes francais*. Paris: Treuttel et Wurtz, 1827.

Von Zeiller, Franz. *Le prescrizioni sul diritto di matrimonio estese al Regno Lombardo-Veneto in virtù della Sovrana Patente 20 aprile 1815 con i commenti a ciascun articolo estratti dal commentario sul codice civile universale della monarchia austriaca*. Milano: Gio. Giuseppe Destefanis, 1815.

Secondary Sources

About, Ilsen, Brown, James, and Lonergan, Gayle, eds. *Identification and Registration Practices in Transnational Perspective: People, Papers and Practices*. Basingstoke: Palgrave Macmillan, 2013.

Agostini, Filiberto. *La riforma napoleonica della chiesa nella repubblica e nel regno d'Italia 1802–1814*. Vicenza: Istituto per le ricerche di storia sociale e religiosa, 1990.

Albergoni, Ginaluca. "La censura in età napoleonica (1802–1814): organizzazione, prerogative e uomini di uno spazio conflittuale." In *Istituzioni e cultura in età napoleonica*, edited by Elena Brambilla, Carlo Capra and Aurora Scotti, 184–219. Milano: Franco Angeli, 2008.

Antonielli, Livio. *I prefetti dell'Italia napoleonica*. Bologna: il Mulino, 1983.

Antonielli, Livio. "L'Italia di Napoleone: tra imposizione e assimilazione di modelli istituzionali." In *Gli imperi dopo l'Impero nell'Europa del XIX secolo*, edited by Marco Bellabarba, Brigitte Mazohl, Reinhard Stauber, Marcello Verga, 409–431. Bologna: il Mulino, 2008.

Baldinelli, Maria Francesca. "Il Consiglio di Stato del Regno d'Italia napoleonico: le procedure e il funzionamento." *Società e storia* 78 (1997): 789–828.

Bernardi, Teresa. "Mobilità e appartenenza multiple a Venezia: il caso di Tarsi alìas Laura Malipiero (1630–1660)." *Genesis*, XVI/2 (2017): 37–59.

Bernardi, Teresa. "Control of Marriage and (Im)mobility in Venice (16th–17th Centuries) in Dialogue with Cultures of Vigilance." *Società e storia* 177 (2022): 546–568.

Bernardi, Teresa and Pompermaier, Matteo. "Hospitality and Registration of Foreigners in Early Modern Venice: The Role of Women within Inns and Lodging Houses." *Gender & History* 31/3 (2019) : 624–645.

Bertrand, Gilles. "L'administration vénitienne et l'évolution des techniques d'enregistrement des étrangers dans le contexte de la Révolution francaise (1789–1797)." *Diasporas. Circulations, migrations, histoire* 29 (2017): 105–129.

Bizzocchi, Roberto. "Marchigiani senza cognome. Un'inchiesta nell'Italia napoleonica." *Quaderni storici* 134 (2010): 533–584.

Bizzocchi, Roberto. *I cognomi degli Italiani. Una storia lunga 1000 anni*. Roma-Bari: Laterza, 2014.

Bortoluzzi, Fabio, "Il precetto politico nelle province venete (1813–1850)." In *Amministrazione della giustizia penale e controllo sociale nel Regno Lombardo-Veneto*, edited by Giovanni Chiodi and Claudio Povolo, 271–289. Sommacampagna: Cierre, 2007.

Brambilla, Elena, "Alfabetismo e società nelle Marche in età napoleonica." In *Istruzione, alfabetismo, scrittura. Saggi di storia dell'alfabetizzazione in Italia (sec. XV–XIX)*, edited by Attilio Bartoli Langeli and Xenio Toscani, 149–185. Milano: Franco Angeli, 1991.

Breckenridge, Keith, and Szreter, Simon. *Registration and Recognition: Documenting the Person in World History*. Oxford: Oxford University Press, 2012.

Brendecke, Arndt. "Attention and Vigilance as Subjects of Historiography. An Introductory Essay." *Storia della Storiografia* 74/2 (2018): 17–27.

Brendecke, Arndt and Molino, Paola. "The Cultures of Vigilance. Historicizing the Role of Private Attention in Society. An Introduction." *Storia della Storiografia* 74/2 (2018): 11–16.

Broers, Michael. "The Parochial Revolution: 1799 and the Counter-Revolution in Italy." *Renaissance and Modern Studies* 33 (1989): 159–173.

Broers, Michael. *The Politics of Religion in Napoleonic Italy: The War Against God, 1801–1814*. London: Routledge, 2002.

Broers, Michael. *The Napoleonic Empire in Italy, 1796–1814. Cultural Imperialism in a European Context?* Basingstoke: Palgrave Macmillan, 2005.

Brubaker, Rogers. *Citizenship and Nationhood in France and Germany*. Cambridge, MA: Harvard University Press, 1992.

Bruley, Yves. "Mariage et famille sous Napoléon. Le droit entre religion et laïcité." *Napoleonica* 14 (2012): 111–126.

Buono, Alessandro. "Le procedure di identificazione come procedure di contestualizzazione. Persone e cose nelle cause per eredità vacanti (Stato di Milano, secc. XVI–XVII)." In *Procedure, metodi, strumenti per l'identificazione delle persone e per il controllo del territorio*, edited by Livio Antonielli, 35–65. Soveria Mannelli: Rubbettino, 2014.

Buono, Alessandro. "Tener persona. Sur l'identité et l'identification dans les sociétés d'Ancien Régime." *Annales. Histoire, Sciences Sociales* 71/1 (2020): 75–111.

Caiani, Ambrogio A. "Collaborators, Collaboration and the Problems of Empire in Napoleonic Italy, the Oppizzoni Affair, 1805–1807." *The Historical Journal* 60/2 (2017): 385–407.

Caiani, Ambrogio A. *To Kidnap a Pope: Napoleon and Pius VII*. New Haven, CT-London: Yale University Press, 2021.

Cambria, Rita. "Brunetti, Vincenzo." In *Dizionario Biografico degli Italiani*, vol. 14 (1972): 585–588.

Canonici, Claudio. "«Per non abbandonare la Chiesa né il popolo». Il giuramento ecclesiastico negli "Stati romani" in epoca napoleonica (1810–1814)." *Rivista di storia del cristianesimo* 1/2 (2004): 303–331.

Canonici, Claudio. "Clero e giuramenti." In *Italia napoleonica. Dizionario critico*, edited by Luigi Mascilli Migliorini, 107–109. Torino: Utet, 2011.

Caplan, Jane and Torpey, John. "Introduction." In *Documenting Individual Identity: The Development of State Practices in the Modern World*, edited by Jane Caplan and John Torpey, 1–12. Princeton, NJ-Oxford: Princeton University Press, 2001.

Cerutti, Simona. "Les "misérables" en droit italien au XVIII e siècle." In *Gens de passage en Méditerranée de l'Antiquité à l'époque moderne. Procédures de contrôle et d'identification*, edited by Claudia Moatti, and Wolfgang Kaiser, 223–242. Paris: Maisonneuve & Larose, 2007.

Cerutti, Simona and Vallerani, Massimo. "Suppliques. Lois et cas dans la normativité de l'époque moderne." *L'atelier du Centre de recherches historiques* 13 (2015). https://journals.openedition.org/acrh/6545

Chevalier, Louis. *Classes laborieuses et classes dangereuses à Paris, pendant la première moitié du XIXe siècle*. Paris: Plon, 1958.

Criscuolo, Vittorio. "Latuada, Felice." In *Dizionario Biografico degli Italiani*, vol. 64 (2005): 44–46.

Dal Cin, Valentina. *Il mondo nuovo. L'élite veneta fra rivoluzione e restaurazione (1797–1815)*. Venezia: Edizioni Ca' Foscari, 2019.

Della Peruta, Franco. *Esercito e società nell'Italia napoleonica: Dalla Cisalpina al Regno d'Italia*. Milano: Franco Angeli, 1988.

Della Peruta, Franco. "Giovanetti, Giacomo." In *Dizionario biografico degli italiani*, vol. 55 (2001): 447–450.

Denis, Vincent. *Une histoire de l'identité: France, 1715–1815*. Seyssel: Champ Vallon, 2008.

Denis, Vincent and Lacour, Pierre-Yves. "La logistique des savoirs. Surabondance d'informations et technologies de papier au XVIIIe siècle." *Genèses* 102 (2016): 107–122.

Desolei, Andrea. *Istituzioni e archivi a Padova nel periodo napoleonico (1797–1813)*. Cargeghe: Editoriale Documenta, 2012.

Di Fiore, Laura. *Alla frontiera. Confini e documenti di identità nel Mezzogiorno continentale preunitario*. Soveria Mannelli: Rubbettino, 2013.

Faron, Olivier and Grange, Cyril. "Une recensement parisien sous la Révolution. L'example des cartes de surete de 1793." *Mélanges de l'Ecole francaise de Rome. Italie et Mediterranée* 2 (1999): 795–826.

Fiocchi Malaspina, Elisabetta. *L'utile giusto. Il binomio economia e diritto per l'avvocato Giacomo Giovanetti (1787–1849)*. Soveria Mannelli: Rubbettino, 2020.

Franzina, Emilio. *Vicenza. Storia di una città. 1404–1866*. Vicenza: Neri Pozza, 1980.

Geselle, Andrea. "Domenica Saba Takes to the Road: Origins and Development of a Modern Passport System in Lombardy-Veneto." In *Documenting Individual Identity: The Development of State Practices in the Modern World*, edited by Jane Caplan and John Torpey, 199–217. Princeton, NJ-Oxford: Princeton University Press, 2001.

Goffman, Erving. *The Presentation of Self in Everyday Life*. Edinburgh: University of Edinburgh, 1956.

Grab, Alexander. "Enlightened Despotism and State Building: The Case of Austrian Lombardy." *Austrian History Yearbook* 19/2 (1984): 43–72.

Grab, Alexander. *Napoleon and the Transformation of Europe*. Bagingstroke-New York: Palgrave MacMillan, 2003.

Grab, Alexander. "Smallpox Vaccination in Napoleonic Italy (1800–1814)." *Napoleonica* 30 (2017): 38–58.

Groebner, Valentin. *Who Are You? Identification, Deception and Surveillance in Early Modern Europe*. New York: Zone, 2007.

Gullino, Giuseppe. "Compagnoni Giuseppe." In *Dizionario biografico degli italiani*, vol. 27 (1982): 654–661.

Higgs, Edward. *Identifying the English: A History of Personal Identification 1500 to the Present*. London/New York: Continuum, 2011.

Hughes, Steven C. *Crime, Disorder and the Risorgimento. The Politics of Policing in Bologna*. Cambridge: Cambridge University Press, 1994.

Ichino Rossi, Costanza. "Della Beretta, Giovanni Antonio." In *Dizionario Biografico degli Italiani*, vol. 36 (1988): 691–694.

Judson, Pieter. *The Habsburg Empire: A New History*. Cambridge, MA-London: The Belknap Press of Harvard University Press, 2016.

Lazzarini, Antonio. "Problemi d'impianto dei servizi demografici in un'area della montagna veneta: il dipartimento della Piave." *Archivio storico di Belluno, Feltre e Cadore* 68/299 (1997): 98–113.

Levati, Stefano. "Società." In *Atlante storico dell'Italia rivoluzionaria e napoleonica*, edited by Maria Pia Donato, David Armando, Massimo Cattaneo, and Jean-François Chauvard, 123–160. Rome: École française de Rome, 2013.

Liva, Antonio. "Gli impiegati." In *La formazione del primo Stato italiano e Milano capitale, 1802–1814*, edited by Adele Robbiati Bianchi, 871–882. Milano: Istituto lombardo Accademia di scienze e lettere, 2006.

Maccabelli, Terenzio. "Melchiorre Gioia e la Statistica del dipartimento del Mella." *Commentari dell'Ateneo di Brescia* 399 (2005): 399–432.

Maitte, Corine. *Les chemins de verre. Les migrations des verriers d'Altare et de Venise (XVIe–XIXe siècles*. Rennes: Presses Universitaries de Rennes, 2009.

Manion, Jane. *Female Husbands: A Trans History*. Cambridge: Cambridge University Press, 2020.

Mazzone, Umberto. "Oppizzoni, Carlo." In *Dizionario biografico degli italiani*, vol. 79 (2013): 384–388.

Meneghetti Casarin, Francesca. *I vagabondi. La società e lo Stato nella Repubblica di Venezia alla fine del '700*. Roma: Jouvence, 1984.

Menozzi, Daniele. "La Chiesa, la rivoluzione francese e l'impero napoleonico." In *Storia della società italiana. L'Italia giacobina e napoleonica*, 143–187. Milano: Nicola Teti, 1985.

Menozzi, Daniele. "L'organizzazione della chiesa italiana in età napoleonica." *Cristianesimo nella storia* 14 (1993): 69–97.

Meriggi, Marco. *Il Regno Lombardo-Veneto*. Torino: Utet, 1987.

Meriggi, Marco. "La cittadinanza di carta." *Storica* 16 (2000): 107–120.

Miccoli, Giovanni. "La Chiesa friulana fra occupazione francese, dominazione austriaca e regno italico." In *Opere d'arte di Venezia in Friuli*, edited by Gilberto Ganzer, 54–63. Udine: Magnus, 1987.

Mori, Simona. *Polizia e statualità nel primo Ottocento: l'esperienza lombardo–veneta e la cultura professionale italiana*. Soveria Mannelli: Rubbettino, 2017.

Naselli, Carmelo Amedeo. *La soppressione napoleonica delle corporazioni religiose. Contributo alla storia religiosa del primo Ottocento italiano*. Roma: Editrice pontificia università gregoriana, 1986.

Noiriel, Gérard. "L'identification des citoyens. Naissance de l'état civil républicain." *Genèses* 13 (1993): 3–28.

Noiriel, Gérard. "Surveiller les déplacements ou identifier les personnes? Contribution à l'histoire du passeport en France de la I[e] à la IIIe République." *Genèses* 30 (1998): 77–100.

Noiriel, Gérard. "Introduction." In *L'identification. Genèse d'un travail d'Etat*, edited by Gérard Noiriel, 16–20. Paris: Belin, 2007.

Oman, Carola. *Napoleon's viceroy, Eugène de Beauharnais*. London: Hodder & Stoughton, 1966.

Pagano, Emanuele. *Enti locali e stato in Italia sotto Napoleone. Repubblica e Regno d'Italia (1802–1814)*. Roma: Carocci, 2007.

Panciera, Walter and Russo, Saverio. "Economia." In *Atlante storico dell'Italia rivoluzionaria e napoleonica*, edited by Maria Pia Donato, David Armando, Massimo Cattaneo, and Jean-François Chauvard, 164–201. Rome: École française de Rome, 2013.

Pasi, Antonia. *Contare gli uomini. Fonti metodi temi di storia demografica*. Milano: LED, 1992.

Pederzani, Ivana. *Un ministero per il culto. Giovanni Bovara e la riforma della Chiesa in età napoleonica*. Milano: Franco Angeli, 2002.

Peruzzi, Piergiorgio. *Progetto e vicende di un Codice Civile della Repubblica Italiana (1802–1805)*. Milano: Giuffrè, 1971.

Pignatelli, Giuseppe. "Codronchi, Antonio." In *Dizionario Biografico degli Italiani*, vol. 26 (1982): 599–604.

Pigni, Emanuele. "La Hoz, Giuseppe." In *Dizionario biografico degli italiani*, vol. 63 (2004): 86–88.

Pillepich, Alain. *Napoleone e gli italiani*. Bologna: il Mulino, 2005.

Pivato, Stefano. *Il nome e la storia. Onomastica e religioni politiche nell'Italia contemporanea*. Bologna: il Mulino, 1999.

Poggi, Stefano, "«Non si tenti un matrimonio fra due donne». "Vero Sesso" e impotenza nell'Italia napoleonica." In *Alla prova delle passioni. Sessualità non conformi e soggettività fra età moderna e contemporane*, edited by Tommaso Scaramella. Pisa: Edizioni ETS, 2024.

Povolo, Claudio. "Dimensioni del sacro e dimensioni del profano. I registri canonici tra istituzioni ecclesiastiche e secolari nella seconda metà del Settecento." In *Archivi e cittadino. Genesi e sviluppo degli attuali sistemi di gestione degli archivi*, edited by Gianni Penzo Doria, 39–72. Sottomarina: Il Leggio, 1999.

Preto, Paolo. "Dandolo, Vincenzo." In *Dizionario Biografico degli Italiani*, vol. 32 (1986): 511–516.

Rao, Anna Maria. *Folle controrivoluzionarie. Le insorgenze popolari nell'Italia giacobina e napoleonica*. Roma: Carocci, 1999.
Rao, Anna Maria. "Napoleonic Italy: Old and New Trends in Historiography" In *Napoleon's Empire: European Politics in Global Perspective*, edited by Ute Planert, 84–97. New York: Palgrave Macmillan, 2015.
Riall, Lucy and Laven, David, eds. *Napoleon's Legacy: Problems of Government in Restoration Europe*. Oxford: Berg, 2000.
Ronsin, Francis. *Le contrat sentimental: débats sur le mariage, l'amour, le divorce, de l'Ancien Régime à la Restauration*. S.l.: Aubier, 1990.
Roscioni, Lisa. *Lo smemorato di Collegno. Storia italiana di un'identità contesa*. Torino: Einaudi, 2007.
Rowe, Michael, ed. *Collaboration and Resistance in Napoleonic Europe: State-Formation in an Age of Upheaval, c. 1800–1815*. Bagingstroke-New York: Palgrave Macmillan, 2003.
Schiaffino, Andrea. "Il ruolo generale della popolazione nell'esperienza storica del Regno Italico." In *Le fonti della demografia storica in Italia. Volume 1. Parte 1*, 519–541. Roma: Comitato italiano per lo studio dei problemi della popolazione, 1971.
Schiaffino, Andrea. "Coscrizione e nuzialità in età napoleonica." In *Reggio e i territori estensi dall'antico regime all'età napoleonica*, edited by Marino Berengo and Sergio Romagnoli, vol. 1, 179–196. Parma: Pratiche, 1979.
Scott, James C. *Seeing Like a State: How Certain Schemes to Improve the Human Condition Have Failed*. New Haven, CT-London: Yale University Press, 1998.
Tassoni, Giovanni, ed. *Arti e tradizioni popolari. Le inchieste napoleoniche sui costumi e le tradizioni nel Regno Italico*. Bellinzona: La Vesconta, 1973.
Torpey, John. *The Invention of the Passport: Surveillance, Citizenship and the State*. Cambridge: Cambridge University Press, 2000.
Torpey, John. "The Rise of States and the Regulation of Movement." In *Procedure, metodi, strumenti per l'identificazione delle persone e per il controllo del territorio*, edited by Livio Antonielli, 185–196. Soveria Mannelli: Rubbettino, 2014.
Toscani, Xenio. "L'alfabetismo nelle campagne dei dipartimenti del Mincio e della Mella e nelle alte valli del Serio e dell'Adda (1806–1810)." In *Istruzione, alfabetismo, scrittura. Saggi di storia dell'alfabetizzazione in Italia (sec. XV–XIX)*, edited by Attilio Bartoli Langeli and Xenio Toscani, 201–244. Milano: FrancoAngeli, 1991.
Vaccari, Andrea. *L'attività del commissario di polizia di Vicenza dal 1807 al 1812*. Parma: Università degli Studi di Parma, 1996–1997.
Viggiano, Alfredo. "Il disordine delle comunità. I giudici di pace napoleonici nei dipartimenti veneti." In *Forme e pratiche di polizia del territorio nell'Ottocento preunitario*, edited by Simona Mori and Leonida Tedoldi, 35–66. Soveria Mannelli: Rubbettino, 2011.
Viggiano, Alfredo. "Carte d'identità nella Venezia Austriaca (1799–1804)." In *Controllare il territorio. Norme, corpi e conflitti tra medioevo e prima guerra mondiale*, edited by Livio Antonielli and Stefano Levati, 363–378. Soveria Mannelli: Rubbettino, 2013.
Viggiano, Alfredo. "Individuare e classificare. Passaporti concessi e negati nella Venezia austriaca del primo Ottocento." In *Procedure, metodi, strumenti per l'identificazione delle persone e per il controllo del territorio*, edited by Livio Antonielli, 105–131. Soveria Mannelli: Rubbettino, 2014.

Viggiano, Alfredo, "Passions déréglées. Race et justice dans la Venise napoléonienne." In *Récit et justice. France, Italie, Espagne, XIVe-XIXe siècles*, edited by Lucien Faggion and Christophe Régina, 243-259. Aix-en-Provence: Presses universitaires de Provence, 2020.

Woolf, Stuart. *Napoleon's Integration of Europe*. London; New York: Routledge, 1991.

Zaghi, Carlo. *L'Italia di Napoleone*. Torino: UTET, 1989.

Zironda, Renato, ed. *Il vicentino tra rivoluzione giacobina ed età napoleonica. 1797–1813*. Vicenza: Biblioteca civica Bertoliana, 1989.

Zovatto, Pietro. "Bressa, Giuseppe Maria." In *Dizionario biografico degli italiani*, vol. 14 (1972): 191–194.

Index

Note: Page numbers followed by "n" refer to end notes.